The Evolution of
Animal Communication

MONOGRAPHS IN
BEHAVIOR AND ECOLOGY

Edited by John R. Krebs and
Tim Clutton-Brock

The Evolution of Animal Communication

Reliability and Deception in Signaling Systems

WILLIAM A. SEARCY

AND

STEPHEN NOWICKI

Princeton University Press
Princeton and Oxford

Published by Princeton University Press, 41 William Street, Princeton, New Jersey 08540

In the United Kingdom: Princeton University Press, 3 Market Place, Woodstock, Oxfordshire OX20 1SY

LIBRARY OF CONGRESS CATALOGING-IN-PUBLICATION DATA

Searcy, William A., 1950–
The evolution of animal communication: reliability
and deception in signaling systems/
William A. Searcy and Stephen Nowicki.
p.   cm.   — (Monographs in behavior and ecology)
Includes bibliographical references and index.
ISBN-13: 978-0-691-07094-0 (cloth : alk. paper)
ISBN-10: 0-691-07094-6 (cloth : alk. paper)—
ISBN-13: 978-0-691-07095-7 (pbk. : alk. paper)
ISBN-10: 0-691-07095-4 (pbk. : alk. paper)
1. Animal communications. I. Nowicki, Stephen, 1955–
II. Title. III. Series.
QL776.S35 2005
591.59–dc22   2004062445

British Library Cataloging-in-Publication Data is available

This book has been composed in Times Roman and Univers Light 45

Printed on acid-free paper. ∞

pup.princeton.edu

Printed in the United States of America

10 9 8 7 6 5 4 3 2 1

We dedicate this book to our families
Margaret, Christopher, Annie
and
Susan and Schuyler

Contents

Figures, Boxes, and Table

Figures

Boxes

Table

Acknowledgments

The first person we want to thank is Peter Marler, who as our postdoctoral advisor taught each of us much of what we know about animal communication. W. A. S. also thanks Gordon Orians, for teaching him how to think in the context of natural selection, and Sievert Rohwer, for discussing a surprising proportion of the issues covered here, too many years ago to be comfortable to remember. S. N. thanks Ben Dane and Ken Armitage for instilling in him his interest in animal behavior and behavioral ecology in the first place, and Tom Eisner for teaching him what it means to ask questions about nature.

In writing this book, we owe a particular debt to Robert Seyfarth and Peter McGregor, who read and commented on the entire text. We also thank Melissa Hughes for reading and commenting on the chapter on aggressive signaling. We are grateful to Sam Elworthy for encouraging us to write the book, and to Sam and Hanne Winarsky for help along the way. For discussion of many of the topics that we cover, we thank Rindy Anderson, Terry Krueger, Sarah Stai, Michael Robinson, Barry Stephenson, Kristy Wolovich, Ana Ibarra, Venetia Briggs, and Kevin Murray at the University of Miami, and present and past members of the Nowicki lab group at Duke University, especially Barbara Ballentine, Martin Beebee, Melissa Hughes, Jeremy Hyman, Sheila Patek, Susan Peters, Jeff Podos, and Denise Pope. For providing unpublished manuscripts, we thank Jane Reid, Jeff Podos, Peter McGregor, Tom Peake, Andrew Terry, Rob Lachlan, Robert Seyfarth, Dorothy Cheney, Julie Gros-Louis, W. Tecumseh Fitch, Clive Catchpole, Kate Buchanan, Karen Spencer, Jill Soha, and Doug Nelson. For help with figures we thank Ken Otter, Melissa Hughes, and especially Mark Mandica, who drew the original illustrations and many of the figures.

We also thank those who have helped us in our own research on animal signaling. We are especially grateful to Susan Peters, who has played a key role in all of this work. We also thank, for their participation in our research, Rindy Anderson, Barb Ballentine, Staffan Bensch, Dennis Hasselquist, Cindy Hogan, Melissa Hughes, Jeremy Hyman, Terry Krueger, Jeff Podos, and Katie Shopoff. We thank the Pymatuning Laboratory of Ecology for logistical support, and the Pennsylvania Game Commission for access to study sites. For financial support, we thank the National Science Foundation and the John Simon Guggenheim Foundation.

The Evolution of
Animal Communication

1 Introduction

Whether signals are reliable or deceptive has been a central question in the study of animal communication in recent years. The crux of the issue is whether animal signals are honest, in the sense of conveying reliable information from signaler to receiver, or deceitful, in the sense of conveying unreliable information, the falsity of which somehow benefits the signaler. This issue arises in a variety of contexts. When a male courts a female, do his signals honestly convey his quality relative to other males? Or does he exaggerate his quality in order to win over females that would otherwise choose some other male? When one animal signals aggressively in a contest over a resource, does the signaler honestly convey its likelihood of attack? Or does the signaler exaggerate that likelihood in order to intimidate competitors that would otherwise defeat him? The question of reliability versus deceit arises even in interactions that, on the face of things, seem to be predominantly cooperative. When an offspring begs for food from its parents, does it honestly convey its level of need? Or does the offspring exaggerate its need in order to get more food than the parents would otherwise provide?

The issue of reliability and deceit in animal communication resonates with human observers for a variety of reasons. One is that the occurrence of deceit is fraught with moral implications. In the view of many, human communication is permeated with deceit. Do humans stand apart in this regard, or are other animals as bad or worse? The answer might have considerable effect on how we view ourselves, as well as on how we view other animals. A second reason for interest in this issue is that the occurrence of deceit, if deceit is defined appropriately, can have considerable implications for our understanding of animal cognition. Some definitions of deceit are framed so as to require cognitive processes of considerable sophistication, such as the ability to form intentions and beliefs and to attribute beliefs to other individuals. If we employ such a definition, and if we can then determine that nonhuman animals deceive each other according to this definition (a big "if"), then we have provided support for a greater level of cognitive capacity than many earlier views of animal behavior have allowed.

Our own interest in reliability and deceit revolves around neither morality nor cognition, but instead derives from the evolutionary implications of the issue. The way one expects animal communication systems to function in terms of reliability and deceit depends on how one views the operation of natural selection. Early students of animal behavior often assumed implicitly that selection operates at the level of groups, so that behavior evolves toward what is best for the population or species as a whole, leading to the view that animal communication consists primarily of the cooperative exchange of reliable information. The predominant view nowadays, however, is that selection acts largely at the level of the individual, so that behavior evolves toward what is best for the individual performing the behavior, and not toward what is best for the group. If behavior is commonly selfish, in this sense, then it is not always obvious why animals should exchange information cooperatively. Instead, one might expect many instances in which signalers would attempt to profit individually by conveying dishonest information. But because individual selection works on the receiver as well as the signaler, receivers ought to respond to signals only if doing so is to their advantage, on average. Therefore, if dishonesty is common, it also is not obvious why receivers should respond to signals.

Taking the argument one step further, if receivers fail to respond to signals, it is not obvious how signaling systems can exist at all. Thus if one accepts the view that selection acts predominantly at the level of the individual, as we do, and if one at the same time accepts the idea that animals do communicate with each other, as seems obvious, then one is left with a series of evolutionary puzzles. Are animal signals in reality reliable or unreliable? If animal signals are reliable, what mechanisms maintain reliability despite the tempting advantages of dishonesty? If animal signals are deceitful, do receivers respond to them anyway, and, if so, why? Our principal purpose in this book is to work through possible answers to evolutionary puzzles such as these.

Definitions

Before we get to these puzzles, we need to define some terms. First, we need to define what we mean by "signal," in order to delimit the set of traits whose honesty and dishonesty we will examine. In one of the first rigorous evolutionary analyses of communication, Otte (1974, p. 385) defined "signals" as "behavioral, physiological, or morphological characteristics fashioned or maintained by natural selection because they convey information to other organisms." Otte explicitly rejected group-selectionist explanations for the evolution of traits, so in his view the transmission of information had to confer some reasonable advantage on the signaler itself in order to satisfy the definition. Thus Otte excluded as signals those traits that convey information to predators or parasites without any benefit to their possessors; he cited the chemicals in human sweat that attract disease-carrying mosquitoes as a possible example. Otte also rejected as signals those traits, such as body size, that may be used by other individuals of the species to assess their possessors but did not evolve for that function. Clearly included under Otte's definition would be vocalizations, color patterns, and body movements that have evolved because they transmit information in a way that benefits the individual that exhibits those traits. More ambiguous are traits, such as the form of a bird's tail, that originally evolved for some other function but have been modified by selection for information transmittal. We will regard such traits, or more precisely their modified properties, as signals; thus the bird's tail itself is not a signal but the tail's length is, if that length has been exaggerated beyond its aerodynamic optimum in order to influence receivers.

This brings us to our definitions of reliability and deceit. In everyday English, "reliable" means that "in which reliance or confidence may be put; trustworthy, safe, sure" (Little et al. 1964). An animal signal, then, would be reliable if one could have confidence in its veracity, or truthfulness—if, that is, one could trust the signal to convey whatever it is supposed to convey. The difficulty with this formulation is in ascertaining what the signal is "supposed to" convey. "Supposed to" in this context must be interpreted from the viewpoint of the receiver rather than the signaler; what matters is whether the signal conveys something that the receiver would benefit from knowing. If we are certain what it is that the receiver benefits from knowing, such as some attribute of the signaler or its environment, then we can ascertain the reliability of the signal by measuring the correlation between the signal and the attribute of interest.

Suppose, for example, that we think that female frogs are interested in the size of conspecific males, and we find that calls communicate information on male size by a negative correlation between call frequency and caller size (males with deeper croaks are larger). We can then determine the reliability of this information by measuring the correlation between call frequency and caller size. The trouble is that we can never really be certain that caller size is what the females "want" or "need" to know. Even if we can show that call frequency is well correlated with caller size, and that the females show a behavioral preference for calls of lower frequency, we cannot be sure that their true interests are not in some other characteristic—perhaps, in this example, male age. The best we can do is to measure as carefully as we can the benefits that the receivers obtain from different types of information. If we can show that female frogs benefit from mating with larger males but not from mating with older ones, we at least can have some confidence that size is what matters to the receivers, and then evaluate reliability of call frequency in terms of its correlation with signaler size.

To formalize this definition, we suggest that an animal signal is reliable if:

1. Some characteristic of the signal (including, perhaps, its presence/absence) is consistently correlated with some attribute of the signaler or its environment; and
2. Receivers benefit from having information about this attribute.

A remaining problem is how to specify what we mean by "consistently correlated." We can never expect a perfect correlation between signal characteristic and the attribute being signaled. Even if signalers are striving for perfect honesty, errors must be expected in the production of the signal and in our measurements of it, either of which would prevent our observing perfect reliability. How good, then, does the correlation have to be for us to conclude that the signal is on the whole reliable? One answer is provided by the concept of "honest on average" (Johnstone and Grafen 1993, Kokko 1997). A signal can be considered honest on average if it contains enough information, sufficiently often, that the receiver on average is better off assessing the signal than ignoring it. Consider again the example of male frogs communicating their size to females via the frequency of their call. The correlation between male size and call frequency can never be expected to be perfect, and in reality is often rather low (see chapter 4). The male's call can be considered honest on average if the correlation between male size and call frequency is good enough that the female benefits on average from using the call to assess male size, instead of ignoring this signal feature. In practice, it will be difficult to determine whether this criterion is being met, but at least it provides a theoretical standard against which reliability can be judged.

A simple way to define "deceptive" would be as the opposite of reliable, but for many the concept of deception carries more baggage, and consequently requires a more complex definition. A relatively simple definition of deception is provided by Mitchell (1986, p. 20), who suggested that deception occurs when:

1. A receiver registers something Y from a signaler;
2. The receiver responds in a way that is appropriate if Y means X; and
3. It is not true here that X is the case.

Note that the definition requires specifying what the signal (Y) means to the receiver. The meaning of Y to the receiver is judged by the response of the receiver to Y together with an observed correlation between Y and X, across many such signals. In other words, we infer that Y means X to the receiver because signalers usually produce Y in association with X, and because the receiver responds to Y in a way that is appropriate if X is true. To make this more concrete, let Y be an alarm call given by the signaler. The alarm call is usually produced when a predator (X) is present, and the receiver typically responds to the alarm call by fleeing, an appropriate (i.e., beneficial) response if a predator is indeed nearby. Deception occurs if the signaler produces the alarm and the receiver reacts by fleeing when in fact no predator is present.

A difficulty with Mitchell's (1986) definition, which he himself points out, is that deception so defined cannot be distinguished from error on the part of the signaler. If the signaler has produced an alarm in error, would we want to call such an action deceptive? This problem can be solved if the definition of deception further stipulates that the signaler benefits from the receiver's response to the signal. Mitchell (1986) himself is uncomfortable with the notion of benefit, remarking that the "idea of benefit is taken from human affairs" and when applied to nonhuman animals typically refers to what a human observer "believes is good for them." For an evolutionary biologist, however, "benefit" has a straightforward meaning—an individual benefits from an action if that action increases the individual's fitness, in the sense of the representation of the individual's genes in subsequent generations. Benefit in this sense is not an anthropocentric idea, but one that applies equally well to all organisms. With the added stipulation about a benefit to the signaler, we will define deception as occurring when:

1. A receiver registers something Y from a signaler;
2. The receiver responds in a way that
    a. benefits the signaler and
    b. is appropriate if Y means X; and
3. It is not true here that X is the case.

Deception defined in this way has sometimes been termed "functional deception" (Hauser 1996), meaning that the behavior has the effects of deception without necessarily having the cognitive underpinnings that we would require of deception in humans.

Other definitions specify that deception must have more complex cognitive underpinnings, that is, that the signaler has an "intention" to cause the receiver to form a false "belief" about the true situation (Russow 1986, Miller and Stiff 1993). Deception defined in this way has been termed "intentional deception" (Hauser 1996). "Intentions" and "beliefs" are mental states, and as such are difficult to measure in nonhuman animals, to say the least. Whether animals possess such mental states, and whether they can ascribe them to others, is of great interest to philosophers (Dennett 1988) and cognitive ethologists (Cheney and Seyfarth 1990, Seyfarth and Cheney 2003, Byrne and Whiten 1992), as well as to the general public. A major goal of some researchers studying deception in nonhuman animals is to use this type of interaction as a window onto the mental states of those animals, in an effort to determine whether they do indeed form intentions, beliefs, and so forth. Although we applaud such efforts, we repeat that our own interests lie elsewhere, in the analysis of reliability and deceit from a functional, evolutionary viewpoint. Another way of saying this is that we are interested in how natural selection shapes animal communication to be either honest or dishonest. From this viewpoint, the question of mental states is largely irrelevant; the costs and benefits to the signaler of giving a false alarm, and to the receiver of responding, ought to be the same whether or not the signaler is able to form an intention and the receiver to form a belief.

Another issue in defining deception is whether to include the withholding of signals. Some authors have argued in favor of this inclusion, suggesting that under certain circumstances, a failure to signal can be considered just as deceptive as producing a dishonest signal (Cheney and Seyfarth 1990, Hauser and Marler 1993a, Hauser 1996). Hauser (1996), for example, states that if an animal fails to produce a signal in a certain context in which that signal is typically produced, and if the animal benefits from failing to signal, that failure constitutes functional deception. This idea seems to us to have little application to a large majority of signaling contexts, such as those involving aggression or mate choice, in which cooperation is not expected from the interactants. In practice, the idea that withholding information is deceptive has most often been applied to cooperative interactions, most notably to interactions in which an animal signals the discovery of a food source to others of the same species (Hauser and Marler 1993a,b). Even here, the concept seems to us to be problematic. Say, for example, that a signaler follows the convention of calling when it finds a large amount of food, more than it can eat itself, and not calling when it finds a smaller amount. The signal then is consistently correlated with an aspect of the environment that receivers benefit from knowing, and so meets our criteria for reliability. Of course the receivers would be even better served by knowing more (i.e., from hearing about the small amount of food as well), but the signaler has not broken its convention in denying them this information.

Before we move on, let us reiterate in less formal terms the definitions of reliability and deceit we plan to use. Reliability requires that there be a correlation between some characteristic of the signal and some attribute of the signaler or its environment that the receiver benefits from knowing about, and that the correlation be good enough that the receiver on average benefits from assessing the signal rather than ignoring it. Deceit requires not only that the correlation between signal characteristic and external attribute be broken at times, but that the signaler benefits from this breakdown. Therefore, if a breakdown occurs in the correlation between signal characteristic and external attribute from which the signaler does not benefit, this would constitute unreliability but not deceit. A breakdown of this type we would describe as "error."

Some History

Opinions about the prevalence of reliability and deceit in animal communication have swung back and forth in recent decades. A convenient place to enter this history is with a seminal paper published by Richard Dawkins and John Krebs in 1978 titled "Animal signals: Information or manipulation?" In writing this paper, Dawkins and Krebs were reacting to what they labeled as the "classical ethological" view of animal communication, which in their opinion treated communication as a cooperative interaction between signaler and receiver. The ethological view assumed that receivers (reactors) were "selected to behave as if predicting the future behaviour" of signalers, while the signalers were "selected to 'inform' reactors of their internal state, to make it easy for reactors to predict their behaviour" (Dawkins and Krebs 1978, p. 289). Thus the classical ethological view held that "it is to the advantage of both parties that signals should be efficient, unambiguous and informative" (Dawkins and Krebs 1978, p. 289). Dawkins and Krebs objected to this Panglossian picture of communication on the grounds that it is not what one would expect to evolve under natural selection. Natural selection favors behavior that enhances the actor's own survival and reproduction, rather than anyone else's, so that "cooperation, if it occurs, should be regarded as something surprising, demanding special explanation, rather than as something automatically to be expected" (Dawkins and Krebs 1978, p. 289). Dawkins and Krebs proposed replacing the cooperative view of communication with one that interprets signaling as an attempt on the part of a signaler to manipulate the behavior of the receiver to the signaler's advantage. Under this alternative, the signaler communicates not in order to tell the receiver what the receiver wants to know, but to induce the receiver to do something that will benefit the signaler. "If information is shared at all it is likely to be false information, but it is probably better to abandon the concept of information altogether" (Dawkins and Krebs 1978, p. 309).

The manipulative interpretation of communication proposed by Dawkins and Krebs reflected the growing consensus among animal behaviorists that individual selection, rather than group selection, plays the preeminent role in shaping the evolution of behavior. Group selection is selection stemming from the births and deaths of groups (such as populations and species) and favoring traits that benefit groups, whereas individual selection is selection stemming from the births and deaths of individuals and favoring traits that benefit individuals. The consensus in favor of individual selection arose in large part in reaction to Wynne-Edwards' (1962) overtly group-selectionist ideas, which brought the distinction between group and individual selection into focus. Group selection as articulated by Wynne-Edwards was sharply criticized, and individual selection championed, by influential evolutionary biologists such as Hamilton (1963), Lack (1966), Williams (1966), and Maynard Smith (1976a). Although argument over these issues has not entirely died away, from the 1970's onward researchers investigating animal behavior have interpreted their results almost exclusively in terms of individual selection.

In describing the "classical ethological" view of communication, Dawkins and Krebs (1978) gave a series of quotations from earlier papers, some of which had a decidedly group-selectionist ring. Tinbergen (1964, p. 206), for example, was quoted as saying "One party—the actor—emits a signal, to which the other party—the reactor—responds in such a way that the welfare of the species is promoted." Ethologists objected that the quotations chosen by Dawkins and Krebs did not correctly represent the central ideas of ethology with respect to communication. Hinde (1981, p. 535), for example, claimed that Dawkins and Krebs had erected a "straw man" that "neither accurately nor adequately conveys the main stream of ethological studies." Hinde (1981) argued that the ethologists, rather than assuming that "signals carry precise information of what the actor will do next," had actually emphasized the use of signals in "conflict" situations, where the animal was torn between "incompatible tendencies" such as attack and retreat. An animal caught in a conflict situation in this way cannot itself predict what it will do next, let alone inform others. And in fact, if one reads the Tinbergen article cited above (Tinbergen 1964), one finds little discussion of either the information transmitted by signals or the selective benefits of signaling to signaler or receiver. Instead, Tinbergen's principal interests were in the evolutionary origin of displays, in the sense of the movements from which signals were originally derived, and in the proximate causation of display, in the sense of what "motivates" the animal to signal. In discussing motivation, Tinbergen (1964) indeed emphasized the "conflict hypothesis," but although this hypothesis may imply that displays have low information content, Tinbergen himself did not draw this inference. As for the level of selection question, Hinde (1981) claimed that the forthrightly group-selectionist statement quoted by Dawkins and Krebs was "not representative" of Tinbergen's writings in general. Hinde (1981) has a point here, in that Tinbergen at times discussed the evolution of behavior in terms of individual as well as group advantage (e.g., Tinbergen 1951), and the same can be said of other "classical" ethologists as well. In truth, Tinbergen's interpretation of the group selection/individual selection distinction was rather different from a contemporary one; for example, he tended to attribute the evolution of any behavior that furthered the reproduction of individuals (rather than their survival) to group advantage rather than individual advantage (Tinbergen 1951).

Whether or not the cooperative information-exchange view of animal communication truly represented the main trend of ethological thinking, this viewpoint was thoroughly discredited by Dawkins and Krebs' analysis. The manipulative view of communication that Dawkins and Krebs suggested in its place had its own problems, however. According to this view, the signaler communicates in order to maneuver the receiver into performing some action that will benefit the signaler, and if the signal can be said to convey any information, that information is at least as likely to be false as true. The critical flaw with this reasoning is that it does not explain why the receiver would be selected to respond to the signal at all. If there is, on average, no information of benefit to the receiver in a signal, then receivers should evolve to ignore that signal. If receivers ignore the signal, then signaling no longer has any benefit to the signaler, and the whole communication system should disappear. In short, it was fairly easy to construct an individual-selectionist argument showing why signaling should not be honest, but when this argument is followed to its logical conclusion, it is not obvious why signaling would occur at all.

A partial solution to this dilemma had already been proposed by Zahavi (1975), with reference to signals used in mate choice. Mating signals provide an excellent example of the honest-signaling dilemma. Females (if they are the ones exercising choice) will benefit from choosing males of superior quality. If superior males can give a signal that identifies them as being superior, then they will benefit from the ability of females to identify and choose them. Poor-quality males, however, would also benefit from being chosen, so they will also be selected to give the signal. If all males, regardless of their quality, give the signal, then the signal contains no information on male quality. Females then should be selected to ignore the signal, and males should cease to give it. The solution proposed by Zahavi was that a mating signal must confer a "handicap" on the survival of the signaler. In Zahavi's words, the handicap serves "as a kind of test imposed on the individual." A male with a highly developed handicap "is an individual which has survived a test" and therefore has demonstrated he is of superior quality. As possible examples of such handicap traits, Zahavi (1975) cited the exaggerated train of the peacock and singing in exposed positions by warblers, both of which he thought would expose a male to predators. "Since good quality birds can take larger risks it is not surprising that sexual displays in many cases evolved to proclaim quality by showing the amount of risk the bird can take and still survive" (Zahavi 1975, p. 211).

Zahavi's handicap idea initially met with skepticism. Maynard Smith (1976b), Bell (1978), and others formulated genetic models to analyze whether a handicap trait would evolve under Zahavi's assumptions. Typically, these models assumed that single genes controlled both the handicap trait in males and the preference for the handicap in females. A third gene controlled viability in both sexes, where viability was defined as fitness exclusive of mating success (Maynard Smith 1985). Maynard Smith (1985) reviewed the results of these modeling efforts, and concluded that with realistic parameter values neither the handicap trait nor the female preference for it would increase from an initially low frequency. A simple quantitative-genetics model by Pomiankowski and Iwasa (1998) recently reinforced these conclusions. The problem is that females benefit from preferring the handicap only to the extent that the preference gene comes to covary with viability, and this covariance arises very indirectly, from the association of the preference with the handicap and the association of the handicap with viability. The resulting weak benefit is counterbalanced by the cost to females of having their sons inherit the handicap, along with its deleterious effect on survival. This cost is sufficient to prevent both the handicap trait from increasing among males and the preference for the handicap from increasing among females (Maynard Smith 1985, Pomiankowski and Iwasa 1998).

The negative conclusions of these genetic models did not lead to the demise of the handicap idea; instead, handicaps were rescued by some new ideas about how the link between a handicap and viability could come about. The original models by Maynard Smith (1976b) and others assumed that the handicap becomes associated with viability because only males of high viability can survive with the handicap. A handicap linked to viability in this way is sometimes referred to as a "pure epistasis handicap" (Maynard Smith 1985), but we prefer the term "Zahavi handicap" (Pomiankowski and Iwasa 1998), both because this label is less cumbersome and because the way this handicap is defined accords well with our reading of Zahavi's original formulation (Zahavi 1975). Following the criticism of his original idea, Zahavi (1977a, p. 603) suggested a new kind of handicap, in which "the phenotypic manifestation of the handicap is adjusted to correlate to the phenotypic quality of the individual." Such an adjustment could be made in a couple of different ways. One is for the handicap trait to be expressed only if a male has both the gene for the handicap trait and the gene or genes for high viability; a handicap of this sort has been termed a "conditional" or "condition-dependent" handicap (West-Eberhard 1979, Andersson 1994). A second way is for the handicap to be expressed in all males that have the handicap gene, but with its size or conspicuousness made to correlate with the viability of its possessor; this has been termed a "revealing" handicap (Maynard Smith 1985, 1991b).

Analysis of genetic models indicated that condition-dependent and revealing handicaps, together with female preferences for them, are much more likely to evolve than are Zahavi handicaps (Andersson 1986, Iwasa et al. 1991). One reason is that whereas all males with the handicap genes pay the full cost of a Zahavi handicap, only a subset of males—those with genes for high viability—pay the full cost of condition-dependent and revealing handicaps (Andersson 1994). A second reason is that the link between handicap and viability is more direct for condition-dependent and revealing handicaps, making them better, more informative signals of viability, and increasing the average benefit of choosing a male with such a handicap (Pomiankowski and Iwasa 1998). Models of condition-dependent and revealing handicaps thus indicated that, given certain assumptions, reliable signals of mate quality might evolve in the context of male courtship of females.

Meanwhile, the reliability of signals given in aggressive contexts was also being debated. In a conflict between two animals, two kinds of information might well be valuable to either contestant: the other animal's "quality," in the sense of its fighting ability or "resource holding potential" (RHP), and its "intentions," in the sense of whether it was likely to attack or to retreat. The near-universal occurrence of display in animal conflicts implied that either or both types of information were commonly being conveyed, but why this was the case was not obvious. In a series of influential papers published in the 1970's, Maynard Smith and colleagues used game theory to analyze this problem (Maynard Smith and Price 1973, Maynard Smith 1974, 1979; Maynard Smith and Parker 1976). We will look at the details of some of these models later (see chapter 4), but the gist of the argument can be presented without mathematics. Suppose that aggressive contests occur between individuals that are well matched for fighting ability, in which case the winner is likely to be the one willing to escalate to a higher level of aggression. If the population had a set of signals that reliably communicated the signaler's aggressive intentions, then it would pay one animal to give way if the other signaled a higher level of aggression than its own. Such a population could be invaded, however, by a cheating strategy whereby the cheater signaled the highest level of aggressiveness, regardless of its true intentions. Cheaters would win many contests without having to fight, and the cheater strategy would increase in frequency. Once cheating became sufficiently common, however, we would expect receivers to evolve to ignore the signal (Maynard Smith and Parker 1976, Maynard Smith 1979). A similar argument could be given for why signals of fighting ability should be dishonest, at least for signals that are not directly constrained by the signaler's phenotype. These theoretical arguments against reliability were widely accepted, and researchers shifted their attention to demonstrating empirically that aggressive displays, on the one hand, did not predict aggression and, on the other, were largely ignored by receivers (Caryl 1979).

Of course if aggressive signals are largely useless in terms of their information content and consequently are ignored by receivers, then it makes little sense for signalers to go on signaling—and yet in practice signaling is nowhere more common than in aggressive conflicts. Here, too, Zahavi's handicap idea, that signals can be honest if they are costly, provided a possible way out of the paradox. Enquist (1985) used game theory to show that reliable signaling of either aggressive intentions or fighting ability could be evolutionarily stable, provided that signaling was costly and that either the cost of signaling or the benefit of winning varied between individuals (see chapter 4). Enquist (1985) acknowledged the importance of Zahavi's (1975, 1977a,b) work in directing attention to the role of signal cost in maintaining signal reliability.

The work of Enquist (1985) on aggressive signals and of Andersson (1986), Pomiankowski (1987), and others on mating signals initiated a swing in opinion back to the expectation that animal signals generally are reliable. This swing was completed by the publication in 1990 of two papers by Alan Grafen. In the first of these, Grafen (1990a) presented a population-genetics model of the evolution of female choice for a male handicap trait. The model assumes that male advertising is costly and that the cost is higher for low-quality than for high-quality males. Given these assumptions, Grafen (1990a) showed that an evolutionary equilibrium exists at which the level of male signaling is a strictly increasing function of male quality—meaning that the signal is reliable—and at which females prefer males with higher levels of signaling—meaning that receivers respond to the trait. In the second paper, Grafen (1990b) presented a game-theory model of honest signaling, which he extended to aggressive signaling as well as mate choice. To obtain an evolutionarily stable strategy, Grafen (1990b) had to assume: (1) that signaling is costly, in the sense that signaler fitness declines as the level of signaling increases; (2) that the receiver's assessment of the signaler's quality increases as the signaler's sig-

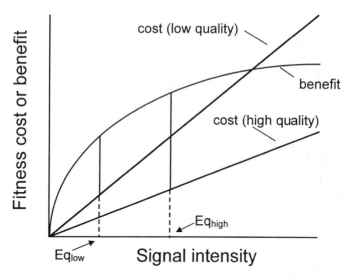

FIGURE 1.1. Johnstone's (1997) first graphical signaling model, in which the signal conveys signaler quality. Two cost lines are drawn, one for a signaler of high quality and a second for a signaler of low quality. The relationship between signal benefit and signal intensity is assumed to be the same for all signalers. The equilibrium signaling level is found as the signal intensity at which the difference between benefit and cost is greatest. The equilibrium for the high-quality signaler (Eq_{high}) is greater than the equilibrium for the low-quality signaler (Eq_{low}).

naling level increases; and (3) that the signaler benefits from being given a higher assessment. In addition, it was necessary to assume (4) that the ratio of the marginal cost of signaling (taken as a positive term) to the marginal benefit of a higher level of assessment is a decreasing function of signaler quality. The latter condition is satisfied if the benefits of improved assessment are the same regardless of quality and the cost of signaling is greater for signalers of poor quality than for those of high quality. Grafen (1990b) also turned the logic around, to argue that the existence of stable signaling systems implies that signals are reliable and costly in a way that meets the above conditions.

We review Grafen's models, as well as other signaling models, in more detail later in the book. For now, we will use graphical models developed by Rufus Johnstone (1997) to aid understanding of the honest-signaling argument. Figure 1.1 shows a version of Johnstone's model that is appropriate for the kind of situation envisioned by Grafen, in which the signal conveys the quality of the signaler, and the signaler benefits from receiving a higher assessment. This benefit increases monotonically with increasing signal intensity, that is, the higher the intensity of the signal, the more effective is the signal in terms of receiver response and the higher is the benefit to the signaler. This version of the model assumes that the benefit of a given signal intensity is the same for all receivers regardless of their intrinsic quality. Signal costs also increase monotonically with signal intensity, but the costs rise more rapidly for a signaler of poor quality than for a signaler of high quality. The optimal signaling level for any signaler occurs at the signal intensity where the difference between benefit and cost is maximized. In order to generate a simple solution, costs are assumed to increase linearly with increasing signal intensity, while benefits increase to an asymptote. Under these assumptions, the optimal signaling intensity is higher for a signaler of high quality than for a signaler of low quality. Thus by assuming that each individual is following its own evolutionary interests, the model generates a signaling system that is reliable, in the sense that levels of signal intensity accurately convey levels of signaler quality.

In this first version of the model, Grafen's fourth assumption concerning the ratio of marginal cost to marginal benefit is satisfied in what has come to be viewed as the standard way—by making the costs of signaling dependent on signaler quality and the benefits independent of quality. Johnstone (1997) uses a second version of the graphical model to show that reliable signaling will occur if these assumptions are reversed. This second version of the model is appropriate for situations in which the signal conveys level of need rather than level of quality; it might be applied, for example, to the case of nestling birds begging for food from their parents. In figure 1.2 the relationship between signal cost and signal intensity is assumed to be the same for all signalers, whereas the benefit of signaling rises more rapidly for a signaler of high need than for a signaler of low need. Again, optimal signaling levels are found where the difference between benefit and cost is maximized for a given signaler, and again the result is reliable signaling, in this case because the signaler with the higher level of need signals at a higher intensity than does the signaler with the lower level of need.

Categories of Signal Costs

Grafen's and Enquist's models convinced many researchers that signal costs can be important in stabilizing signaling systems. Attention then turned to determining whether signals actually do have costs, and how those costs might come about. Many categories of costs have been described, but it is important to note that all must be reducible to a fitness cost if they are to be effective in enforcing signal reliability. This does not necessarily mean that a signaler's fitness is lower the higher its level of signaling, because (among other considerations) it might receive fitness benefits from the responses that receivers make to its signals. A precise definition of cost was provided by Grafen (1990b), who considered a signal to be costly if the partial derivative of fitness with respect to signaling level was negative, holding receiver assessment and sig-

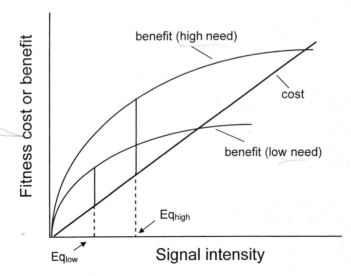

FIGURE 1.2. A second version of Johnstone's (1997) graphical model, in which the signal conveys signaler need. Two benefit curves are drawn, such that the signaler with higher need receives a greater benefit than the signaler with lower need, at any given signaling level. The relationship between signal cost and signal intensity is assumed to be the same for all signalers. The equilibrium signaling levels (compare figure 1.1) are again found as the signal intensities at which the difference between signal benefit and signal cost is greatest. The signals are predicted to be reliable, in the sense that the signaler with the higher need signals at higher intensity than does the signaler with lower need.

naler quality constant. In other words, the concept of signal costs requires that signaler fitness go down as signaling goes up when any fitness effects of receiver response and of signaler quality are held constant.

In categorizing costs, a primary division can be made between so-called "receiver-dependent costs" and "receiver-independent costs"—costs that stem from some response of receivers to a signal versus costs that are imposed regardless of whether or how receivers respond (Guilford and Dawkins 1995, Vehrencamp 2000). Vehrencamp (2000) suggests further dividing the receiver-dependent category into "vulnerability costs" and "receiver retaliation rules." A vulnerability cost occurs because the action of producing the signal opens the signaler to an increased chance of injury, if a receiver chooses to attack. Zahavi (1987) gave putative examples of signals in this category, such as postures and vocalizations that require relaxation and are given as aggressive displays. Zahavi argued that relaxation in the proximity of an aggressive rival was dangerous, and that the risk would be less for an individual of strong fighting ability than for one of poor fighting ability. The basis of a receiver-retaliation rule is that receivers are more likely to attack, or punish in some

other way, those signalers that give one kind of signal rather than an
Enquist (1985) used a receiver-retaliation rule as the cost of an aggre
signal in one of his original signaling models: one of two signals is more
effective than the other, in terms of helping the signaler to win the contest, but
is also more likely to provoke the opponent to fight. Receiver-retaliation rules
are most likely to apply to aggressive signals, but they can apply to mating
signals, if rivals of the same sex are more likely to attack a signaler if it gives
a more effective mating signal than if it gives a less effective one.

We consider receiver-independent costs to include three categories: production costs, developmental costs, and maintenance costs. Production costs are
costs that are paid at the time the signal is exhibited to the receiver. Included
in this category would be the considerable energy consumed by calling in frogs
or roaring in red deer, the time taken away from foraging and other activities
by singing in a songbird, and the increased risk of predation a male stickleback
may experience when it exposes its red coloration. Developmental costs are
costs paid at the time a signal develops, well before the signal is displayed.
The concept of developmental costs is usually applied to display structures
whose growth requires considerable investment, such as the antlers of deer
(Andersson 1986). We have argued that certain display behaviors also have
developmental costs, especially complex behaviors that are supported by specialized neural systems (Nowicki et al. 1998, 2002a). Maintenance costs are
ones that are a consequence of having to bear a display structure once it has
been developed, and which are paid regardless of whether the display is actually given. A prime example of this category is the cost paid by birds for an
elongated tail. Elongation of tail feathers beyond a certain point makes flight
more clumsy and more expensive (Evans and Thomas 1992), so a male with
an elongated tail is likely to expend more energy in flying, to be more vulnerable to predation, and to have decreased foraging success—all of which are
detrimental to fitness.

Some displays must have multiple costs. The extravagant train of the peacock provides a familiar example. Growing these greatly elongated feathers
must require a considerable investment in energy and nutrients, a clear instance
of developmental costs. Fanning the train to display it must require some energy expenditure, and may also expose the signaler to an increased risk of
predation, both of which are production costs. Displaying the train might have
some receiver-dependent costs, if a large display tends to elicit attacks from
rival males. And as the peacock's train must have an aerodynamic impact,
maintenance costs due to decreased flight performance certainly apply. Measuring the summed effects of all these different costs, with their different units
(energy versus risk) and timing, would be very difficult indeed.

In the real world, signals often attenuate and/or degrade while propagating
between signaler and receiver, making it difficult for receivers to discriminate
signals from irrelevant energy (Wiley 1994). Given this problem, and assuming

that a response in the absence of a signal is costly, it follows that receivers may be selected to set a high threshold for response in order to avoid "false alarms" (Wiley 1994, Johnstone 1998). If receiver thresholds are high, signalers will be selected to produce intense and hence costly signals in order to ensure detection (Johnstone 1998). By this argument, signals may be costly for reasons of "efficacy" rather than reliability. To support the handicap principle, one needs to show that a signal has a "strategic cost" over and above its "efficacy cost" (Maynard Smith and Harper 2003); however, in real-world signaling systems the boundary between strategic and efficacy costs may be difficult or impossible to delineate.

Alternative Explanations for Reliability

The handicap principle—the idea that signals can be reliable if they are costly in an appropriate way—is not the only viable explanation for honesty in animal signals. We have already introduced a second explanation, that embodied in Johnstone's (1997) second graphical mode (figure 1.2). In that model, it is the relationship between signal intensity and benefit that chiefly acts to make signaling honest, rather than the relationship between signal intensity and cost. If the benefit of signaling is sufficiently different for various categories of signalers, then optimal signaling levels will be quite different for different signalers, even if signaling costs are minimal.

A third explanation for reliability is the lack of a conflict of interest: if signaler and receiver agree on the rank order of possible outcomes of their interaction, then signals can be reliable without being costly. We introduce this idea at greater length in chapter 2. A fourth explanation, based on a model by Silk et al. (2000), is that deceit can be disadvantageous if receivers remember acts of deception by particular signalers and discriminate against signals from those individuals in the future. We call this mechanism "individually directed skepticism." Again, this hypothesis is explained more fully in chapter 2.

A fifth explanation for reliability is that some signals are constrained to be honest because of the mechanisms by which they are produced. Maynard Smith and Harper (2003) term a signal that is constrained to reliability in this way an "index," which they define as "a signal whose intensity is causally related to the quality being signalled, and which cannot be faked." To give a concrete example, the fundamental frequency of a vocalization might be considered to be an index of body size. The argument is that the fundamental frequency is determined primarily by the size of the vocal-production apparatus, for example by the length of the vocal folds (or vocal "cords") in many vertebrates. Longer cords produce lower frequencies, vocal-cord length is correlated with body size, and therefore small animals are constrained to produce higher frequencies than large animals.

Note that this argument requires two assumptions: first, that there is an inherent relationship between a signal property and the structure that produces the signal, and second, that there is some necessary relationship between the structure that produces the signal and an attribute of the signaler of interest to the receiver. We consider the validity of these assumptions with respect to vocal signals and body size in chapter 4. For now, we want to point out that an argument based on these sorts of constraints can often be recast in terms of developmental costs. In the example just described, the size of the vocal apparatus, and thus the length of the vocal cords, is not absolutely determined by body size; rather, the size of the vocal apparatus can vary independently of overall size to some degree. It is possible, then, for an individual to develop a vocal apparatus of larger size than that of other individuals of identical body size, but it might pay various developmental costs for doing so. Thus the boundary between handicap and index signals is not always clear.

Deception Redux

Grafen's (1990a, b) models in particular were tremendously influential in convincing researchers that signal reliability is not only possible but probable. Suddenly, the major theoretical puzzle was not how signals could possibly be reliable, but how they could ever be deceptive. Grafen (1990b) himself was moved to ask "What happened to cheating?" As an example of a scenario that might allow deception to occur, Grafen (1990b) suggested that two groups of signalers might exist, for one of which signaling was cheaper than the other, holding signaler quality constant. Those for whom signaling was less costly would signal at a higher level than expected and would benefit from the discrepancy, fulfilling our criteria for deception. Another possibility is that some signalers (for whatever reason) receive greater benefits from signaling than others do and signal more intensely than expected for that reason. More complicated scenarios are also possible, in which both the costs and benefits of signaling differ across signalers. Costs and benefits might differ with respect to signaler age, history, physiological state, and so forth.

In subsequent chapters we will review models in which the balance between costs and benefits makes deception an evolutionarily stable strategy, or ESS, within stable signaling systems. In such "ESS models" (Grafen and Johnstone 1993), a signaling system with some admixture of deception can only be stable if receiver response to the signal is adaptive on average. This condition does not necessarily require that deceptive signals be rare, or even in the minority; deception can be the rule rather than the exception if the benefit of responding to an honest signal is high enough and the cost of responding to a false signal is sufficiently low.

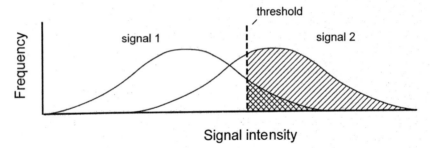

FIGURE 1.3. The problem of discriminating between two signals that vary along a single dimension, in this case signal intensity (adapted from Wiley 1994). Signal 1 is given by signalers of low quality; signal 2 is given by signalers of high quality. The receiver sets a threshold and responds only to signals that exceed that threshold. Those signals above the threshold and under the signal 2 curve represent correct detections (hatching with positive slope); those above the threshold under the signal 1 curve represent false alarms (hatching with negative slope). Sliding the threshold to the right decreases the number of false alarms but at the cost of decreasing the number of correct detections as well. Conversely, sliding the threshold to the left increases the number of correct detections but at the cost of increasing the number of false alarms as well. It is impossible to set the threshold at a value that simultaneously minimizes false alarms and maximizes correct detections.

Another way to view deception is as a consequence of failure on the part of receivers to discriminate between classes of signals; this viewpoint puts deception into the realm of signal detection theory (Wiley 1994, Getty 1995, 1996). Suppose we have two classes of signalers differing in some attribute important to receivers, such as quality. The two classes produce signals that differ along a single dimension, such as intensity (figure 1.3). Signals vary within a class, and there is some overlap between the two classes. A receiver that benefits from discriminating in favor of the signalers of high quality can set a threshold below which it will not respond and above which it will. A receiver that sets a high threshold minimizes its chances of responding to signals from low-quality individuals (i.e., it minimizes false alarms), but at the cost of failing to respond to signals from some high-quality individuals (i.e., of missing correct detections). Thus, no threshold can simultaneously minimize false alarms and maximize correct detections. The optimal receiver strategy depends on the frequency of the two signaler classes and the payoffs of the various possible outcomes (false alarms, correct detections, etc.) (Wiley 1994). If the cost of a false alarm is low and the benefit of a correct detection high, then "adaptive gullibility" may be favored, with the threshold set low so that a receiver often responds to incorrect signals (Wiley 1994). Those incorrect signals that are above threshold can be considered to deceive the receiver.

The signal-detection and ESS approaches to deception are not antithetical. To simplify somewhat, the signal-detection approach assumes that the correct and incorrect signals cannot be completely separated by a receiver, and receiver response is maintained because the benefits of responding to the subset of correct signals above the optimal threshold outweigh the cost of responding to the subset of incorrect signals above that threshold. The ESS approach assumes that the correct and incorrect signals are not separable at all, and receiver response is maintained because the benefits of responding to the complete set of correct signals outweigh the costs of responding to the complete set of incorrect signals. The two approaches can be combined, by introducing signal variation into ESS models (Johnstone 1998), but most ESS models do not make this step. For many of the actual signaling systems that we will examine, honest and deceptive signals appear to be identical, so that a signal-detection approach does not apply. One could still argue, however, that the honest and dishonest signals might be separable on the basis of context.

Dawkins and Guilford (1991) have argued that less-reliable conventional signals will replace honest-handicap signals because of the costs that handicap signals impose on receivers. Costs to receivers are particularly likely in systems, such as roaring contests between red deer (Clutton-Brock and Albon 1979), where one individual can induce another to signal maximally only by signaling maximally himself. A more general cost to receivers is the time investment that in many cases is necessary to assess displays; for example, a female songbird might have to listen to a male for a considerable period in order to estimate the size of his song repertoire. Time spent attending to a display imposes an opportunity cost, in the sense that the receiver might be investing its time in something else of value, and may also impose a survival cost, if proximity to a signaler increases risk of predation. Dawkins and Guilford (1991) argue that, given these and other possible costs of receiving elaborate signals, it would often be advantageous to both receiver and signaler for the signaler to use cheap conventional signals instead of handicaps.

The term "conventional signal" has been used with various meanings in the animal-communication literature (e.g., Maynard Smith and Harper 1988, Dawkins and Guilford 1991), but the clearest definition in our opinion is that provided by Guilford and Dawkins in a later paper (Guilford and Dawkins 1995). For them, a conventional signal is one in which there is "a degree of arbitrariness in the relationship between signal design and signal message" and therefore a need for an agreement (a convention) on what the signal means (Guilford and Dawkins 1995, p. 1692). A nonarbitrary signal might be one in which the signal has a cost that is related to the message, as when the "signal 'uses up' some of the quality being signalled about" (Guilford and Dawkins 1995). For a conventional signal there is no such relationship between signal design, signal cost, and signal message. Instead, the costs of conventional signals are inherently receiver-dependent, and the convention of what the signal means is maintained solely by the response of the receivers.

Implicit in Dawkins and Guilford's (1991) argument, then, is the assumption that receiver-dependent costs are less effective in maintaining signal reliability than are other types of costs. As we discuss later (chapter 4), both reliability and deceit can emerge from signaling models in which the only signal costs are receiver-dependent. The same is true, however, of models in which signals have production, developmental, or maintenance costs (chapter 3); that is, these models too can generate reliable signals but also support some level of dishonesty. Thus it is not obviously true that receiver-dependent costs are more likely to allow dishonesty in signaling systems than are other categories of cost. At the same time, it does seem to be true that theory allows some level of dishonesty in signaling systems of most types.

Evolutionary Interests of Signalers and Receivers

Honest signaling in the absence of signal costs would be expected if the signaler and receiver have identical interests in an evolutionary sense, meaning that a fitness gain experienced by one individual produces an equal fitness benefit for the other. Communication between two such individuals would be akin to communication between two cells or two organs within an individual, and one in general would not find reliability puzzling for signaling systems that operate within individuals (but see Zahavi and Zahavi 1997, who suggest that costs are important in ensuring the reliability of signaling between cells within single individuals). Evolutionary interests also ought to be identical for separate individuals if they are genetically identical.

More commonly, communication occurs between genetically distinct individuals, and here we can distinguish three likely cases with respect to evolutionary interests. In the first, the interests of signaler and receiver are overlapping though not identical. This description is most likely to apply when signaler and receiver are genetic relatives. For related individuals, a fitness benefit experienced by one is necessarily experienced by the other as well, though to a reduced degree. Conflicts of interest are still possible, whereby the optimal outcome of an interaction differs for the two interactants. In chapter 2, we discuss theoretical models that have been proposed for cost-free, reliable signaling between individuals with overlapping interests. We then examine what is known empirically about two of the best-studied examples of communication between relatives: begging, in which offspring solicit food or some other resource from their parents, and alarm signaling, in which one individual warns another of the approach of a predator. Alarm signaling can be interpreted in ways other than as signaling between relatives, for example as signaling between prey and predator. We consider these possible interpretations before examining aspects of reliability and deceit in this type of system. Finally, in chapter 2 we introduce another explanation for reliability, which can be thought of either as existing outside of the costly signaling paradigm or as simply positing a novel kind of cost. Here reliability is advantageous to a signaler because it interacts multiple times with a receiver able to recognize signalers as individuals and to remember their past record of reliability (Silk et al. 2000). In this situation, even if the signaler gains a benefit from deceiving a receiver on one interaction, that benefit may be outweighed by the cost of having the receiver fail to respond to its signals during subsequent interactions.

A second possibility is for the evolutionary interests of signaler and receiver to be separate but not necessarily opposing. We describe this situation as "divergent interests," and consider it to apply to mate attraction and mate choice. Suppose a male acts as the signaler, trying to attract a female, who acts as the receiver. Signaler and receiver in this context typically are genetically unrelated, so genetic relatedness does not provide a tie between their respective fitnesses. A large body of both empirical evidence and theory suggests that females will benefit from assessing prospective mates on some aspect or aspects of "quality." Females therefore will be selected to attend to signals of male quality, as long as those signals are reliable. Males, however, might benefit from exaggerating their quality, other things being equal, thus pushing the signals toward unreliability. This kind of system was the principal focus of Grafen's (1990a,b) signaling models. We review those models in greater detail in chapter 3, together with more recent models that attempt to show how deception might coexist with reliability in mate-attraction signals. We then review some of the empirical results on reliability and deceit in mating signals. We do not attempt a comprehensive review of the vast literature on male attributes that affect female choice of mates (Andersson 1994). Instead, we focus on just three categories of mating signals, chosen because they illustrate particularly well the issues of reliability and deceit. These systems are carotenoid signals in fish and birds, song in songbirds, and tail length in birds.

A third possibility is for the evolutionary interests of signaler and receiver to be in direct opposition. This description applies most generally to cases in which two animals engage in an aggressive contest for possession of some resource, such as food, territory, or mates. Here the interests of the two interactants are necessarily opposed, in the sense that if one animal wins the resource the other loses it. Signaling models, starting with that of Enquist (1985), have suggested that reliable signaling can occur in aggressive interactions despite the opposing interests of signaler and receiver. We review those models in chapter 4, and then examine some actual aggressive signaling systems. Again, we focus on just a subset of the systems that have been studied, chosen because they illustrate the issues of interest. The aggressive signaling systems we have chosen to review are postural displays and badges of status in birds, weapon displays in crustaceans, and calling in frogs and toads.

In many signaling interactions, there are individuals other than the primary interactants that benefit from acquiring whatever information is exchanged.

When individuals that are not directly involved in a signaling interaction nevertheless gather information from it, their behavior is termed "eavesdropping" (McGregor 1993, McGregor and Dabelsteen 1996). Eavesdropping may have important implications for the reliability of a signal, in the sense that unintended receivers can impose additional costs on a signaler (or, in theory at least, may provide additional benefits), and thus a signaler may be selected to modify the reliability of signals that are subject to eavesdropping. Signaling very commonly occurs in "networks" of signalers, rather than in closed dyadic interactions, suggesting that eavesdropping may often be a key factor in the evolution of signaling systems. In chapter 5, we discuss evidence that eavesdropping occurs, and we explore the implications of eavesdropping for signal reliability.

In examining each of the signaling systems that we review in chapters 2, 3, and 4, we will take a standard approach, one that is based on the logic of the reliable signaling problem. One way of stating this logic is that the existence of a signaling system, in which a signaler signals and a receiver responds, implies that the signal has some appreciable level of reliability. The reliability of the signal in turn implies, according to theory, that the signal has some appreciable cost. Therefore, in addressing each system we begin by reviewing evidence on whether receivers actually respond to the signal in question. Demonstrating that receivers respond is the crucial step in establishing the existence of a signaling system. Once a signaling system has been shown to exist, the next step is to examine whether the signal is reliable, as predicted by the logic sketched out above. Reliability can be established by assessing the correlation between attributes of the signal and whatever it is that the receiver benefits from knowing. If the signal is indeed reliable, we next assess the signal's costs. Unfortunately, theory is not terribly clear on the magnitude of the costs needed to maintain reliability; nevertheless, we can make some headway in determining whether the signal is more costly than seems needed simply for transmission. Finally, for each signaling system we review, we will discuss any evidence adduced for the deceptive use of the signal. Certain of the systems that we will focus on have been chosen because they do provide convincing evidence of deception.

Throughout we focus on natural signaling systems, rather than on signaling systems imposed on animals by humans. This focus is in keeping with our evolutionary interests; we want to see what signaling systems natural selection has come up with, not what humans can induce animals to do. We also will concentrate, although not quite exclusively, on communication within species. We have made this choice because it is in the within-species context that there is a clear contrast between the older view of communication as essentially cooperative and the newer view of communication as the product of selection for behavior that furthers each individual's own interests. This decision causes us to exclude some classic cases of deception, such as Lloyd's (1965, 1986) demonstration that females of the predatory firefly genus *Photuris* mimic the flash patterns of female fireflies of the genus *Photinus*, and in this way lure male *Photinus* to their deaths. Although fascinating, in our view such examples of interspecific deception do not pose the same kind of evolutionary puzzles as intraspecific deception, because no one would expect communication across species to be cooperative. Where we do discuss interspecific communication is in cases where there is debate about whether a signal has evolved for a within-species or between-species function.

Another decision we have made is not to make use of one-time observations of the behavior of individual animals. Such observations have long been used as evidence for the occurrence of deception in nonhuman animals, dating back at least to the work of Romanes (1883). Romanes (1883) gathered examples of deception from lay observers as part of his attempt to establish the occurrence of intelligent behavior in animals, and then interpreted those examples rather liberally ("Another of my correspondents, after giving several examples of the display of hypocrisy of a King Charles spaniel . . ."). In more modern analyses, one-time observations of apparent deception, made by scientifically trained observers, have been systematically collected and categorized for primates by Whiten and Byrne (1988, Byrne and Whiten 1992), again with an eye chiefly to the implications of this type of evidence for animal intelligence. As our interests are not in the cognitive aspects of deception, we can avoid anecdotal evidence without passing any general judgment on its scientific usefulness (Burghardt 1988, Byrne and Whiten 1988). We confine our attention to signaling interactions that occur regularly enough that they can be statistically analyzed using data from single studies. We emphasize experimental evidence whenever possible, but note that experimental methods are more applicable to some of the questions we address (such as the response of receivers) than to others. In particular, the reliability of signals often can be addressed only by measuring correlations between the attributes of signal and signaler.

2

Signaling When Interests Overlap

The interests of two individuals overlap in an evolutionary sense when the fitness of one depends, at least in part, on the fitness of the other. Such a positive fitness relation occurs whenever two individuals are genetically related; because they share genes, the overall success of one relative's genes depends to some extent on the success of the other's. Additional causes of convergent interests are possible, for example when the members of a mated pair depend on each other's continued survival and good health for successful reproduction, or when the members of a group depend on each other for safety from predation or to obtain food or other resources. Genetic relatedness, however, is the cause of overlapping interests that has been most emphasized in signaling research.

We begin by considering theory, which in practice means models of signaling between relatives. These models provide two basic explanations for signal reliability: it may result from the absence of conflict of interest between signaler and receiver, or it may be maintained by signal costs. We then consider empirical evidence from studies of three rather different types of signals: solicitations, alarms, and food calls. In solicitation, or "begging," one individual appears to ask another for some resource. Success in soliciting resources would seem to require overlapping evolutionary interests, and in practice successful solicitation most often occurs between genetic relatives, especially between offspring and their parents. Alarms are signals in which one individual warns others of the approach of some danger, such as a predator. One explanation for the occurrence of alarm signals is that they evolve to aid genetic relatives, although (as we will see) this explanation is controversial. We next consider the information available on food calls, calls that are given when food is discovered and that often have the effect of recruiting others to the food source. Food calls in some cases are given between individuals with overlapping interests, but they are also given between individuals with diverging and opposing interests. Our discussion of food calls leads us to consider another type of explanation for the maintenance of signal reliability: memory by receivers of the past performance of individual signalers with whom they interact multiple times.

Signaling Between Relatives: Theory

Much of the theoretical work on signaling between relatives can be traced back to a short paper in which John Maynard Smith (1991a) introduced the "Sir

Philip Sidney game." Sir Philip Sidney was an English poet and soldier of the sixteenth century who fought in the Netherlands during the Dutch rebellions against Spain. In a story familiar to all British schoolboys of an earlier generation, Sir Philip, having been wounded at the battle of Zutphen, is supposed to have given his water bottle to a dying soldier, saying "Thy necessity is yet greater than mine." According to Maynard Smith (1991a), this "unusual example of altruism by a member of the English upper classes" was the inspiration for his model.

What characterizes the Sir Philip Sidney (SPS) game in general is that a signaler solicits a potential donor for a resource. The signaler is either needy or not needy, and, just as important, its condition cannot be assessed directly by the donor. The survival probability of the signaler will be raised if the donor gives it the resource, and this benefit is greater for a needy signaler than for one not needy. Because giving the resource lowers the survival of the donor, selection normally will not favor giving. If signaler and donor are related, however, giving may be favored, especially if the signaler is needy. To allow for the handicap argument, in which signal costs impose reliability, the signal can be given a cost, such that producing the signal itself lowers the signaler's chance of survival. A signaling system will exist if at equilibrium the signaler is more likely to signal if needy than if not, and if the donor is more likely to give the resource if the signal is produced than if it is not.

The original SPS model of Maynard Smith (1991a) is simple, and its mathematics consequently are unusually transparent. Furthermore, the model illustrates a number of important conclusions. We therefore reproduce the model in some detail in box 2.1. What characterizes this particular model is that the signal is discrete, thus conveying that the signaler is either needy or not needy; varying levels of need cannot be conveyed. In addition, because a single donor interacts with a single signaler, the donor does not have to choose between competing signalers. One important conclusion to emerge from this model is that if there is any conflict of interest between signaler and donor, the signaling system can be stable only if the signal has a cost. A conflict of interest in general means that selection on one interactant's genes would favor a different outcome than selection on the other's genes. In this case, a conflict arises if the signaler benefits from a transfer of resources even when not needy, while the donor benefits from a transfer only if the signaler is needy. The obverse conclusion also holds: cost-free signaling is possible if there is no conflict. For example, it is possible for both interactants to benefit from a transfer of resources when the signaler is needy, and for neither to benefit when the signaler is not needy; it should be intuitively satisfying that cost-free signaling can be evolutionarily stable under such circumstances.

A variety of complicating factors can be added to this simple SPS model to give it greater realism. Following Godfray (1991) and Johnstone and Grafen (1992), the first complication we consider is to make the signaling system continuous rather than discrete. Johnstone and Grafen's (1992) model is closest to

Box 2.1.
THE SIR PHILIP SIDNEY GAME OF MAYNARD SMITH (1991A)

The signaler's chance of survival is assumed to depend simultaneously on his need and on whether he receives the needed resource, as follows:

signaler needy/gets resource: survival = 1
signaler needy/does not get resource: survival = 0
signaler not needy/gets resource: survival = 1
signaler not needy/does not get resource: survival = V, where $0 < V < 1$.

In addition, the signaler's chance of survival is mutliplied by $1 - t$ if he signals; thus t is the cost of signaling, and can be set at 0 for cost-free signaling.

The donor's chance of survival depends on whether he gives up the resource, as follows:

donor keeps resource: survival = 1.
donor gives resource: survival = S, where $S < 1$.

With these assumptions, Maynard Smith (1991a) proceeds to give the fitnesses of each interactant, depending on the strategy followed in the interaction. The fitness (W) of either interactant is the individual's own probability of survival plus r times the survival of the other, where r is the degree of relatedness of the two. The possible strategies for the donor are: D0, give only in response to the signal; Dm1, always give; Dm2, never give. If p is the probability that the signaler is needy, and if it is assumed that the signal is reliable, then:

$$W(D0) = (1 - p)(1 + rV) + p[S + r(1 - t)]$$
$$W(Dm1) = (1 - p)(S + r) + p[S + r(1 - t)]$$
$$W(Dm2) = (1 - p)(1 + rV) + p[1 + r(0)] = (1 - p)(1 + rV) + p$$

D(0) is stable if it is superior to either alternative. $W(D0) > W(Dm1)$ reduces to:

$$1 + rV > S + r \tag{1a}$$

$W(D0) > W(Dm2)$ reduces to:

$$S + r(1 - t) > 1 \tag{1b}$$

Three strategies also are possible for the signaler: B0, signal only if needy; Bm1, always signal; and Bm2, never signal. Assuming that the donor only gives the resource if the signal is received, the fitnesses of the signaler strategies are:

$$W(B0) = (1 - p)(V + r) + p(1 - t + rS)$$
$$W(Bm1) = (1 - p)(1 - t + rS) + p(1 - t + rS)$$
$$W(Bm2) = (1 - p)(V + r) + pr$$

Again, for the honest-signaling strategy B(0) to be stable it must be superior to either alternative. $W(B0) > W(Bm1)$ reduces to:

$$V + r > 1 - t + rS \tag{2a}$$

$W(B0) > W(Bm2)$ reduces to:

$$1 - t + rS > r \tag{2b}$$

(Box 2.1 continued)

Receiving the resources will be beneficial to the signaler, even if it is not needy, as long as the benefit to its own survival $(1 - V)$ is greater than the cost to the survival of the donor $(1 - S)$ discounted by r, in other words as long as:

$$1 - V > r(1 - S) \qquad (3a)$$

It may benefit the donor to give the resource even when the signaler is not needy, if the benefit to the signaler discounted by r is greater than the cost to the donor, in other words if:

$$r(1 - V) > 1 - S \qquad (3b)$$

This merely rewrites the condition for 'always give' to be favored (1a reversed), and if 'always give' is favored, then signals are not needed. More interesting is the case where giving when the signaler is not needy is against the donor's interests, or:

$$1 - S > r(1 - V) \qquad (3c)$$

Eqs. 3a and 3c together define the conditions for a 'conflict of interests': transfer of the resources is always good for the signaler, but is only good for the donor if the receiver is needy. 3a and 3c cannot be simultaneously true if $r = 1$, but can be true with realistic values of r such as 0.5.

Condition 3a can be rewritten as $V + r < 1 + rS$. This and condition 2b can be simultaneously true only if $t > 0$, in other words if there is a signal cost.

If the 'always give' strategy is favored in the donor, then there is no conflict of interests but no signaling system is needed. There can, however, be conditions under which no conflict exists and yet signaling is favored. Suppose that the signaler does not want the resource if he is not needy, meaning that:

$$1 - V < r(1 - S) \qquad (4a)$$

At the same time, the donor does not benefit from transferring the resource if the signaler is not needy:

$$1 - S > r(1 - V) \qquad (4b)$$

Under these circumstances, both parties benefit from a reliable signal. Maynard Smith (1991a) gives an example that satisfies conditions 1, 2 and 4: $r = 0.5$, $S = 0.8$, $V = 0.95$, and $t = 0$. Note that, in this example, the signaling system is stable with no signal cost.

To conclude, if there is a conflict of interest the signal system can be stable only if the signal has a cost. Without a conflict of interest, cost-free signaling can be favored.

Abstracted with permission from J. Maynard Smith. 1991. Honest signalling: The Sir Philip Sidney game. *Animal Behaviour* 42:1034–1035. Elsevier. ∎

the original SPS game. As in the original, the resource is unitary, and is either donated or not as a whole. Need is made continuous by allowing y, the survival probability of the signaler without the resource, to take any value between 0 and 1. Signal intensity is also made continuous. The signaler's strategy is a function, Q(y), that determines the signal intensity as a function of y. Signal intensity is measured by the extent to which the signal reduces the signaler's survival, which makes intensity directly proportional to cost. The donor is assumed to give up the resource if and only if it receives a signal above some threshold p. The donor's strategy is a function P(x), which determines p as a function of the donor's survival probability (x) without the resource. The survival probabilities, x and y, are assumed to have uniform distributions between 0 and 1. The survival probability of the signaler becomes 1 if it obtains the resource; therefore the benefit of successful signaling (the difference in survival with and without the resource) increases as need (the probability of mortality without the resource) increases. Inclusive fitness for either player depends on its own survival plus the survival of the other, discounted by their relatedness (r). Johnstone and Grafen (1992) solve for this model's equilibrium conditions, at which neither player can better its inclusive fitness by an incremental change in behavior. At equilibrium, the signal is reliable, in the sense that it accurately reflects need, and it is costly, in the sense that it lowers the signaler's fitness.

Godfray (1991) considers the case of a single offspring soliciting resources from a single parent. As in Johnstone and Grafen (1992), both need and signal intensity are assumed to be continuous functions, and in addition the amount of resource delivered is also allowed to vary continuously. Delivery of resources to the offspring is assumed to increase its fitness but with diminishing returns; thus, for example, the delivery of a particular amount of food has a greater effect on the fitness of a hungry offspring than on that of a satiated one. Because delivery of resources by a parent is assumed to diminish its future reproduction, the parent must trade the fitness of the present offspring against the production of future offspring. Thus, an offspring must trade the benefit it obtains from signaling, which stems from the signal's effect on increasing resource delivery by the parent, against two costs, a direct cost of signal production and an indirect cost due to a decrease in the number of its future sibs that the parent will produce.

Godfray's (1991) model embodies these and some additional assumptions, for example that signal cost is a linear function of signal level. He then solves for the equilibrium at which neither party benefits from an incremental and unilateral change in behavior. Figure 2.1 shows one solution. Note that the level of signaling increases monotonically as condition decreases (moving from right to left in fig. 2.1), so in that sense the signal is reliable. And because resource delivery by the parent increases with increasing signal intensity, the signal is effective. Godfray (1991) concludes that a costly signal of need between relatives can be reliable.

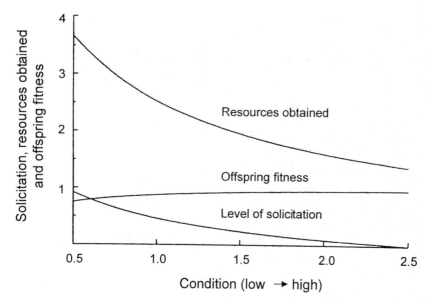

FIGURE 2.1. A solution of Godfray's (1991) model of the signaling of need by a single off-spring to its parents. In the evolutionarily stable strategy (ESS), the signal is reliable, in the sense that the level of solicitation increases monotonically as the offspring's condition decreases (i.e., as its need increases). Also, in the ESS, the receivers respond to the signal, in the sense that the resources obtained by the offspring from the parents increase in parallel with the level of solicitation. Poor condition negatively affects offspring fitness, but that effect is largely (but not entirely) counterbalanced by the higher level of parental investment in offspring of low condition. (Reprinted with permission from H. C. J. Godfray. 1991. Signalling of need by offspring to their parents. *Nature* 352:328–330. Nature Publishing Group.)

Thus far, the models seem to support the possibility that begging systems can be reliable, but a substantial problem lurks: the costs to both signaler and donor may be high enough at the signaling equilibrium that both parties would be better off without the signaling system. Using Godfray's (1991) model, Rodríguez-Gironés et al. (1996) show that an alternative equilibrium exists at which the parent delivers a fixed amount of resources to its offspring without regard to the offspring's behavior, and the offspring conserves energy by not signaling. These authors then compare the fitness of both parent and young at this nonsignaling equilibrium to their fitness with signaling. The result depends on the assumed probability distribution for offspring condition, but with most distributions both parent and young have higher expected fitness at the nonsig-naling equilibrium than they do with signaling. Bergstrom and Lachmann (1997) come to the same conclusion for both the discrete and continuous Sir Philip Sidney games: at the stable signaling equilibrium, "both players may be worse off than they would be with no signalling at all." If signaling is so costly, then we have reason to doubt whether evolution would reach the signaling equilibrium or remain there.

The models we have examined so far have concentrated on the question of how much resource a donor should deliver to a single recipient. In real begging systems there often may be more than a single offspring in a brood, raising the question of how food should be allocated among individuals when there are several possible recipients. As a first step in addressing this problem, Godfray (1995) models the case in which two offspring beg for resources from their parents. Godfray assumes that the parents have a fixed amount of resource to deliver and therefore do not work harder as the level of begging increases. This assumption eliminates any cost of resource delivery stemming from a decrease in future reproduction. Instead, the only cost parents experience when they allocate more resource to one offspring is that incurred because the other current offspring receives less. As in Godfray's (1991) single-offspring model, the fitness of an offspring is assumed to increase as it receives more resources, but with diminishing returns. Specifically, Godfray (1995) assumes that the fitness benefit, f, is related to the amount of food obtained, y, by the function

$$f = B(1 - e^{-cy})$$

where B is the asymptotic benefit (using the notation of Royle et al. 2002) and c is a rate constant. Assuming that $B = 1$, figure 2.2a shows the relationship between fitness benefit and resource share for different values of c. Clearly, the benefit approaches the asymptote more rapidly with higher values of c. Godfray (1995) interprets the rate constant c as representing offspring condition, the rationale being that an offspring in good condition needs less food to raise its fitness to the asymptotic value than is needed by a chick in poor condition. Signal costs are assumed to increase linearly with the begging level, x. The parents monitor the begging levels of their two chicks and adjust their resource allocation such that the fixed amount of food they have is entirely allocated and the marginal fitness gains of the two offspring are equalized. Each chick also monitors the other's begging level, in order to assess the other's condition.

As with the models we discussed previously, Godfray (1995) concludes that an evolutionarily stable signaling system is possible as long as the signal is costly. At equilibrium, the predicted signaling level declines as the signaler's condition increases (figure 2.2b). If need is assumed to be inversely proportional to condition, then this relationship implies that signaling level increases with need, and the signal is in that sense reliable. The condition of the nestmate, however, also affects signaling level; one offspring of fixed condition will signal more intensely as its nestmate's condition worsens. Another prediction is that the level of signaling should be lower for higher levels of relatedness between nestmates; this occurs because one of the costs to exaggerating one's own need is a decrease in resources delivered to one's nestmate.

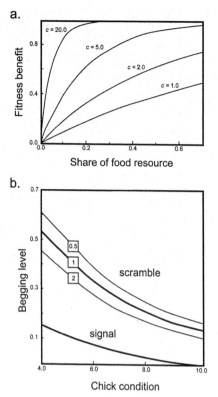

FIGURE 2.2. The assumptions and results of models in which multiple offspring within a brood beg for food from their parents. **a**. The relationship between the fitness benefit accrued by a chick and the share of the parent's food delivery it obtains, as assumed in Godfray (1995) and Parker et al. (2002). The form of the relationship is $f = B(1 - e^{-cy})$, where f is the fitness benefit, B is the asymptotic benefit (here assumed to be 1), c is a rate constant, and y is the food share. Curves are shown for four values of c. From Parker et al. (2002). **b**. Relationships between the ESS level of begging and the condition of the focal chick (A) predicted by the honest-signaling model of Godfray (1995) and by the scramble model of Parker et al. (2002). Three curves are shown for the scramble model, for three different ratios of chick competitive abilities (a/b): 0.5 (chick A is the weaker competitor), 1 (equal competitors), and 2.0 (chick A is the stronger competitor). Condition of chick B is set at 7. (Reprinted with permission from G. A. Parker et al. 2002. Begging scrambles with unequal chicks: Interactions between need and competitive ability. *Ecology Letters* 5:206–215. Blackwell Publishing.)

Johnstone (1999) further extends the behavioral validity of begging models by adding two factors not included in Godfray's (1995) model: he allows the number of competing offspring to be greater than two, and he allows signaling to affect the total amount of resources delivered as well as their allocation. Surprisingly, he finds that as the number of competitors increases, the level of signaling tends to decrease, except at the highest level of need. Johnstone (1999) suggests that the decrease in signaling intensity occurs because with many competitors, the majority of individuals give up, because any modest increase in their own signal level will still not allow them to compete successfully with the neediest members of the brood. The addition of multiple signalers therefore alleviates the unprofitability-of-signaling problem; with two or more competitors, and a reasonable level of relatedness between signaler and donor (that between parent and offspring) and between competing signalers (that between full sibs), signaling is more profitable than is nonsignaling, at least to the signaler.

Lachmann and Bergstrom (1998) explore a different approach to the too-costly-signaling dilemma, by considering solutions involving partial signaling. Most continuous begging models seek "separating" equilibria, in which any two signalers of differing need give signals of differing intensities. Lachmann and Bergstrom (1998), using the Sir Philip Sidney game, show that alternative, "pooling" equilibria exist, in which signalers of different need sometimes give identical signals. Bergstrom and Lachmann (1998) go on to show that pooling equilibria can be cost-free; for example, if both the neediest and least-needy signalers would rather reveal their need than be perceived as having average need, then a signaling equilibrium exists in which the signalers divide themselves into two pools with no cost. For reasonable levels of relatedness, both signalers and donors have higher fitness at this equilibrium with two signaling pools than at the no-signaling equilibrium. Equilibria are possible with more pools, but additional pools are of little advantage to either signaler or donor. Cost-free signaling is stable in a two-pool system only if the ratio of maximum need to minimum need exceeds a critical value that increases as r, the relatedness between signaler and donor, decreases (Brilot and Johnstone 2003).

What happens at the two-pool equilibrium is that signalers divide into two groups based on their condition: those below a certain threshold signal that they are needy, and those above the threshold signal that they are not (presumably by not signaling at all). Further gradations of need are not signaled. The system can be stable despite a certain amount of conflict of interest, for the following reason. Signalers of some intermediate range of condition, below the threshold for signaling, might benefit from receiving the resource from donors whose own condition is sufficiently high. To get this class of donors to donate, these signalers would have to give the signal of need. Signaling need would induce additional donors, of lower condition, also to donate, and having these additional donors donate would not be in the interests of the signalers, because of kinship effects. The benefit of deceiving the former class of donors is balanced by the cost of deceiving the latter class, and the signalers below the threshold therefore do not signal (Lachmann and Bergstrom 1998).

Johnstone (2004) has further investigated the effects of nestmates on each other's begging. In Godfray's (1995) model, the begging of one chick always increases as the condition of its nestmate deteriorates. Johnstone (2004) argues that this outcome results from a simplifying assumption made by Godfray: that begging affects only the allocation of food among the chicks and not the total amount of food delivered by the parent. If begging affects the total food delivered, then begging can be in part cooperative, in the sense that one chick benefits if another chick begs intensely and induces the parents to deliver more total food to the brood. Johnstone (2004) models this possibility by assuming two components of begging, a competitive component that affects allocation, and a cooperative component that affects the total amount of food delivered. At equilibrium, the competitive component always increases with decreasing nestmate condition, but the cooperative component may remain unchanged or even decrease. Whether begging can ever be separated into distinct competitive and cooperative components is unclear to us, however. If Johnstone (2004) allows a single aspect of begging to affect both allocation and total deliveries, then begging always increases as nestmate's condition worsens, as in Godfray's (1995) original model.

Another modeling tradition exists, apart from the "honest-signaling" models that we have discussed so far. In this other tradition, begging is viewed as an element of "scramble competition" through which nestlings vie with one another for access to parental investment (Royle et al. 2002). A fundamental difference between the two traditions is that in scramble models the offspring control allocation of resources, whereas in honest signaling the parents have control. Early scramble models did not address questions of reliability, in the sense that the condition of the offspring was not included as a parameter, and the models could thus not evaluate the relationship between signaler condition and signal intensity (Macnair and Parker 1979, Parker and Macnair 1979). Recent scramble models, however, have considered signaler condition (Rodríguez-Gironés et al. 2001a, Parker et al. 2002). Parker et al. (2002), for example, propose a scramble model that parallels the honest-signaling model of Godfray (1995), in which two chicks compete for a fixed amount of resources delivered by their parents. Rather than solving for the pattern of allocation that is best for the parents, Parker et al. (2002) assume the parents allocate in proportion to the relative begging intensities of the two chicks; this embodies the passive-allocation assumption of the scramble tradition. Begging intensities as assessed by the parents are modified by the competitive abilities of the chicks. Thus if we let x_A = the begging level of nestling A, and x_B = the begging level of nestling B; and furthermore if a = the competitive ability of A, and b = the competitive ability of B; then the share of the total resources that is obtained by A is $(ax_A/(ax_A + bx_B))$.

Parker et al. (2002) assume that the cost of begging increases linearly with the level of begging, and that the benefit has the same asymptotic relationship with begging level as shown in figure 2.2a. The condition of the chick is again represented by c, the parameter that controls the rate at which the benefit curve approaches the asymptote. As Royle et al. (2002) point out, this and similar scramble models make many of the same predictions made by honest-signaling models (Kilner and Johnstone 1997); for example, parents are predicted to respond to begging in their provisioning, and begging is predicted to be costly. In addition, begging is predicted to be reliable, at least in a certain sense. Figure 2.2.b compares the predicted relationships between the ESS levels of begging and the condition of the begging chick for parallel parameter values of the Parker et al. (2002) scramble model and the Godfray (1995) honest-signaling model. For the scramble model, three curves are shown, each for a different ratio of the competitive ability of the focal chick to the competitive ability of its sib (a/b). In all cases, the begging level decreases monotonically as chick condition increases.

If we assume that the rate constant c can be interpreted as condition, that condition is negatively related to need, and that need is what parents want to know, then the above analysis tells us that begging can be reliable under the assumptions of both honest-signaling and scramble models. There are, however, more complex ways of interpreting need. Royle et al. (2002), for example, suggest that need can be defined as the amount of increase in fitness that an offspring experiences per unit of food obtained. This definition may correspond less well with a commonsense idea of need than does the simpler, reciprocal-of-condition definition, but it arguably corresponds better to what the parents would benefit from knowing. The fitness benefit per unit food is given by the slope of the benefit curves in figure 2.2a. Note that at a high level of food allocation, the slope of the benefit curve is higher for the chick of low condition (c = 1.0) than for the chick of high condition (c = 20.0), but at low food allocation the slope is higher for the chick of high condition. Thus at high food, need (defined as fitness benefit per unit food) is inversely proportional to condition, as we have previously assumed, but at low food the relationship is reversed, and need is directly proportional to condition (Parker et al. 2002). Given the more complex definition of need, then, a signal that is always honest about condition is not always honest about need.

This problem forces us to reexamine the benefit function (illustrated in figure 2.2a) that is assumed by both Godfray (1995) and Parker et al. (2002). An attractive feature of this function is that the benefit of being fed at first increases rapidly per unit food obtained by the chick, with the increase in benefit smoothly declining to 0 as the chick becomes satiated. Another attractive feature is that the parameter c, interpreted as condition, could be varied to obtain curves that reach the asymptotic fitness level with less food for a chick in good condition than for a chick in poor condition. The features of this benefit function are not, however, entirely what we would like. Let us interpret condition, for the moment, as being determined solely by hunger, or more specifically by time since last feeding. The fitness function that we have been using then assumes that a chick that has recently been fed (e.g., c = 20 in figure 2.2a) starts at the same fitness as a chick that has not been fed in a very long time (e.g., $c = 1.0$). Instead, it seems more reasonable to suppose that the former, almost-satiated chick would start at a higher fitness level than the latter, almost-starving chick, so that the overall benefit of a feeding would be less for the former than for the latter. To solve this problem, we could model hunger with a single one of the fitness curves in figure 2.2a, say, the one for $c = 5$. A chick that has just been fed would start at the upper right of the curve, and then descend down to the left along the curve as time since the last feeding increases. A chick at the upper right would have high condition/low hunger; a chick at the lower left would have low condition/high hunger. The attractive feature of this way of modeling condition is that any signal that was honest about need in the sense of hunger would also be honest about need in the sense of the increase in fitness per unit food obtained, because the two are strictly proportional.

Both the honest-signaling model and the scramble model suggest that a signal can be reliable about need, at least in the limited sense of level of need defined as hunger. Both predict that parents will respond to begging, and both include costs of begging as a necessary assumption for evolutionary stability of the signaling system. Both sets of models thus conform in many ways to the basic handicap assumptions and predictions. The scramble models predict higher signaling costs than do the honest-signaling models, which may be the best way of testing between the two.

Although models of begging have been moving toward greater complexity and thus greater realism, they still fall short of the complexities of nature in many respects. One rather alarming oversimplification has been pointed out by Rodríguez-Gironés et al. (1998): all the models we have discussed consider a single interaction between signaler and donor, rather than the iterated interactions common in nature, especially in the case of offspring interacting with parents. We expect that incorporating iterated interactions may further alleviate the problem of too-costly signaling, by making the inefficiencies of the nonsignaling alternative more apparent. If offspring do not signal need, then the best a parent can do is to deliver the amount of food needed by an offspring of average condition. For a parent feeding a single offspring over an extended period, this strategy would require that the parent be able to keep a running tally of past food deliveries. For a parent feeding multiple offspring, the nonsignaling strategy would require that the parent be able to identify individual offspring and to keep a running tally of deliveries to each. Two parents feeding multiple offspring would each have to track the other parent's deliveries as well as its own. To the extent that parents are incapable of such feats, the signaling alternatives become relatively more attractive.

While waiting for more realistic models to appear, we can draw certain provisional conclusions from existing models. Signals can be reliable when animals solicit resources from related individuals. Reliable signaling between relatives can be cost-free, especially in cases of partial signaling, where only

FɪɢᴜʀE 2.3. Stages in the begging intensity of nestling magpies (from Redondo and Castro 1992). As intensity increases, new components are added: raising the neck, vocalizations (represented by the balloons), standing, and wing movements. Individual components can also increase in intensity, for example, vocalizations can increase in rate and amplitude. (Reprinted with permission from T. Redondo and F. Castro 1992. Signalling of nutritional need by magpie nestlings. *Ethology* 92:193–204. Blackwell Publishing.)

limited information is communicated. The reliability of more informative, continuous signals passed between relatives can be maintained if the signals are costly, as proposed by the handicap mechanism. In general, the costs needed to maintain reliability go down as the relatedness between signaler and receiver goes up. Finally, although models predict that costly signals will be reliable for the most part, they still allow room for dishonesty in certain senses, as when one offspring increases its own signaling intensity in response to an increase, not in its own need, but in the need of a competing signaler.

Begging

When a parent bird arrives at the nest, the nestlings perform a graded series of behaviors collectively known as begging. In the European magpie (*Pica pica*), for example, a nestling begs at low intensity simply by raising its head slightly and gaping (figure 2.3). At higher intensity, the nestling produces vocalizations ("begging calls") while it gapes. Next, the nestling stands and stretches its head upward, again gaping and vocalizing. Finally, at highest intensity, the nestling stands, stretches upward, gapes, and vocalizes while flapping its rudimentary wings (Redondo and Castro 1992). This pattern is typical of begging birds. Not only are components added one after another to produce a graded series of intensity levels, but individual components, such as the amplitude and rate of the begging calls, are themselves graded in intensity. The behavior of the chick certainly gives the impression that it is asking for, or demanding, food from its parent, especially as all these behaviors will subside if the nestling receives a meal. The interaction thus matches closely that modeled in the Sir Philip Sidney game, and (as we have seen) many of the models of signaling between relatives are explicitly based on this kind of interaction.

In examining empirical studies of begging, we use the series of steps outlined in chapter 1. First, we consider whether receivers respond to the signal—that is, whether the begging of nestlings causes parents to increase their delivery of food. Second, we ask whether the signal is reliable, that is, whether the signal conveys to the parents what the parents need to know, which is assumed to be their offspring's level of need. Third, we discuss whether begging is costly in a way that acts to ensure honesty. Fourth, we consider evidence for deceit, in the sense of offspring exaggerating their need for their own benefit. Finally, we refer to the models discussed in the preceding section, to ask whether the predictions of the models are upheld in a way that convinces us that the models adequately explain how this communication system functions.

RESPONSE OF PARENTS TO BEGGING

Nonexperimental evidence on parental response to begging is often ambiguous. In Wilson's storm-petrels (*Oceanites oceanicus*), for example, parents bring food to a single young, which is sequestered in a nest burrow. Quillfeldt (2002) showed that the number of begging calls produced by an offspring during a parental visit was positively associated with the amount of food that the parent delivered. This result is consistent with the inference that parents respond to the begging of their offspring, but it is equally compatible with the interpretation that offspring change their begging in response to cues that parents give indicating their willingness to feed. Young grey seals (*Halichoerus grypus*) use begging calls to solicit feeding from their mothers, in a system similar to the begging of nestling birds. Smiseth and Lorentsen (2001) found that mothers approached and offered to suckle their single offspring in 85% of the 15-minute periods following begging, compared to only 6% of randomly chosen 15-minute periods. This pattern again is consistent with the parent having responded to the offspring's solicitation, but it could also be explained if both the offspring's probability of begging and the mother's probability of feeding increase with time since last feeding.

Studies of species with multiple young per brood have measured whether the amount of food that parents allocate to specific chicks in a nest is correlated with the chicks' absolute or relative begging intensities. This approach usually requires videotaping the parental feeding visits, as otherwise it is impossible to determine accurately which chick is fed and how intensely it has begged. Using this method, Smith and Montgomerie (1991a) showed that in American robins (*Turdus migratorius*) the chick fed first during a parental visit on average starts to beg earlier, extends its neck higher, and holds its bill closer to the parent than do broodmates that are not first to be fed. In broods of two young in pigeons (*Columba livia*), the chick that begs longer receives 70 to 80% of feedings early in the nestling phase, although this percentage declines toward 50% as the chicks grow older (Mondloch 1995). These results are consistent with parents responding to begging, but again other mechanisms are possible. One alternative is that parents rotate feedings among their young, so that each is fed at the same average rate over the course of several feedings. If chicks beg more as their hunger level rises, then feeding rotation also predicts a correlation between begging intensity and feeding allocation. A second alternative is that parents receive some cue to nestling hunger that is independent of begging, such as gut distension. Again, even if parents ignored begging, the use of an independent cue to hunger would produce a correlation between begging and food allocation, as long as chicks beg more the hungrier they are. Neither of these two alternative mechanisms can explain the correlation found by Stamps et al. (1989) in budgerigars (*Melopsittacus undulatus*) between average begging intensity measured over several days and average feeding rate over the same period. This result, however, can be explained if both feeding rate and begging intensity depend on some other, long-term attribute of individual young, such as sex or order of hatching.

To test among these possible mechanisms, experiments are needed in which begging intensity is manipulated independently of hunger and other attributes of the young. Two of the common experimental approaches, depriving young of food or giving them supplemental feedings, clearly do not achieve the desired independence between begging and need. A number of other experimental designs remain, however, some of them quite ingenious.

One of the first experiments to manipulate begging cues was performed by von Haartman (1953) with pied flycatchers (*Ficedula hypoleuca*). Von Haartman used a nestbox divided into two compartments, one closed and one open. The open compartment contained a single nestling being cared for by its parents. Additional nestlings could be placed in the second compartment, allowing them to be heard but not seen from the first compartment. This arrangement allows manipulation of the auditory component of begging but not of the visual. In three trials, parents brought food an average of 2.4 times per half hour when the second compartment was empty, and 7.1 times per half hour when the second compartment contained six begging nestlings.

A more modern version of von Haartman's experiment is to use playback of recorded begging calls to supplement the auditory component of begging. Several studies have used this design, mainly with birds (Muller and Smith 1978, Bengtsson and Rydén 1983, Harris 1983, Davies et al. 1998) but also with domestic pigs (*Sus scrofa*) (Weary and Fraser 1995). We will use a trio of recent studies to illustrate the range of responses that have been observed. Two of the three studies were done with red-winged blackbirds (*Agelaius phoeniceus*), both of them using playback of begging calls recorded from 3-day-old broods, and both playing the calls from speakers hidden near the nests.

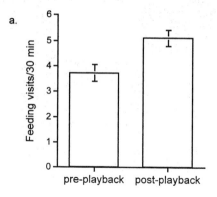

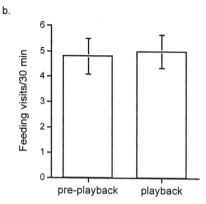

FIGURE 2.4. The response of female red-winged blackbirds in two experiments in which nestling begging calls were played at nests. **a.** Burford et al. (1998) found a significant increase in the number of feeding visits made by the female parent during the 30 minutes after playback of five minutes of begging vocalizations, relative to the 30 minutes before. **b.** Clark and Lee (1998) found no increase in the number of feeding visits made by the female parent during the first 30 minutes of playback of begging vocalizations, relative to the 30 minutes before.

Despite considerable similarities in methods, the results of these two studies differed markedly. In a sample of 30 broods, Burford et al. (1998) found that the number of feeding visits by females was 37% higher during the 30 minutes after 5 minutes of begging-call playback than during the 30 minutes before playback; this increase was highly significant (figure 2.4a). By contrast, in a sample of ten broods, Clark and Lee (1998) found that female feeding visits increased by a mere 4% during the first 30 minutes of 2 hours of playback, a nonsignificant change (figure 2.4b). One explanation for the difference in results is that the females in the Clark and Lee (1998) experiment were working near maximum before playback, and thus were capable of little increase in effort, whereas the females in Burford et al. (1998) were working less before playback (figure 2.4a) and thus were better able to increase their effort. At any rate, the positive results of Burford et al. (1998) are more typical of begging-call playback studies in general, at least of published studies, than are the negative results of Clark and Lee (1998).

Price (1998) played begging calls for 2 hours at nests of yellow-headed blackbirds (*Xanthocephalus xanthocephalus*). Rates of parental feeding visits doubled during playback, compared to control periods. An interesting feature of this study is that Price (1998) also measured effects on nestling weight gain, finding that gains were significantly higher during playback than during control periods. In a second experiment, Price (1998) broadcast begging calls near randomly chosen nests over a 5-day period, and found significantly greater weight gain over this longer period in the experimental broods than in the controls. This result indicates that parents can be stimulated to an increased effort over an extended period.

One criticism of these playback studies is that the nestlings under observation may increase their own begging in response to playback of begging calls, and the parents may be responding to the increase in nestling begging rather than to playback. Playback of begging calls has been shown to stimulate increased begging in nestlings of some species (Muller and Smith 1978) but not others (Kilner et al. 1999). Even if the parents are responding to increased begging of their own nestlings rather than directly to playback, these experiments still show that parents respond to some aspect of increased begging by increasing their rate of feeding. Moreover, parents have been shown to respond to playback of recorded begging calls in the absence of any of their offspring, for example in domestic pigs (Weary et al. 1996).

In order to manipulate all aspects of begging, rather than just begging calls, Litovich and Power (1992) fed nestling European starlings (*Sturnus vulgaris*) with alcohol-soaked raisins, enough to calm the activity of the young without putting them to sleep. Parental feeding of drugged nestlings was compared to feeding of the same nestlings after they had been fed raisins that had not been soaked in alcohol. Broods were fed significantly less often when drugged than when not drugged, presumably at least in part because drugging lowers their level of signaling.

Thus far we have discussed the posturing and calling that young perform in what can be termed "active begging." Kilner (1997) studied a rather different kind of begging signal in canaries (*Serinus canaria*). When a very young canary nestling opens its mouth to beg, the color inside its mouth changes rapidly from brown to bright red. Kilner (1997) manipulated mouth color in canary chicks by adding red dye to their food; this treatment intensified the red color in the mouth for about 1 hour. Kilner (1997) gave chicks an equal amount of either colored or uncolored food, and then observed feeding by the parents over the next 30 minutes. Parents delivered more food to broods of two when both chicks had reddened mouths than when neither had reddened mouths. When one chick within a brood was reddened and the other was not, parents allocated food preferentially to the reddened chick. Similar experiments manipulating nestling mouth color have been done in two other species, great tits (*Parus major*) (Götmark and Ahlström 1997) and barn swallows (*Hirundo rustica*) (Saino et al. 2000), and in both cases the parents again increased food deliveries to the chicks whose mouths were reddened.

A variety of evidence, both observational and experimental, converges to show that receivers in begging systems do respond to the signals given by their young. Specifically, parents respond to higher levels of active begging and to redder mouth color by increasing their provisioning of the young. Thus receivers not only respond, they respond in a way that is beneficial to the signalers.

RELIABILITY OF BEGGING

In considering the reliability of active begging, empirical researchers have assumed that what the signal conveys to the receiver is the signaler's need. This is just the assumption made in most models of begging, but the empiricists have added a distinction not often drawn by the theoreticians, the distinction between short-term and long-term need (Price et al. 1996). Short-term need is synonymous with hunger, and can be operationally defined as the amount of food required to satiate an individual. Long-term need refers to the total amount of resources that a given offspring will require from its present age until independence. Long-term need may depend on offspring gender, for example if males are destined to be larger than females at independence. Long-term need also may depend on offspring condition, in the sense of mass corrected for linear dimensions; a chick whose mass is low for its size may need extra food in order to avoid starvation and to achieve an optimal asymptotic weight. Long- and short-term need can vary independently; for example, a male nestling of low mass that has just been fed would have a high long-term need and a low short-term need. Note that the need modeled by Godfray's (1995) fitness curves (figure 2.2b) is solely a short-term need, which can be satisfied by a single feeding.

A strong case can be made that active begging reliably signals short-term need. Some of the evidence is correlational. Redondo and Castro (1992) showed that food intake by nestling magpies, measured as the proportion of body weight added in 1-hour periods, was negatively correlated with subsequent begging intensity; in other words, chicks with low recent food intake begged more intensely. Reliability in signaling short-term need can also be demonstrated by simple experiments in which hunger is manipulated either by artificial feeding or by short-term deprivation. For example, Cotton et al. (1996) brought European starling chicks into the lab, where they were fed

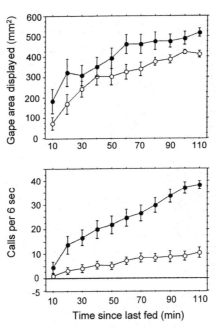

FIGURE 2.5. The relationship between begging intensity and the duration of food deprivation in broods of four reed warbler chicks (from Kilner et al. 1999). The chicks were fed to satiation and then deprived of food for 110 minutes. The upper graph shows how the total gape area exposed by all four chicks increases with time since feeding; the lower graph shows the change in the rate of begging calls. Open circles are means (± se) for 3–4-day-old chicks; closed circles are means for 6–7-day-old chicks. Food deprivation has a highly significant effect on both displays.

either one, two, or three food items every 30 minutes for 4 hours. The chicks then were stimulated to beg and their responses were videorecorded. Time spent begging decreased dramatically as the level of feeding increased. In the converse experiment, Kilner et al. (1999) withheld food from broods of four reed warbler (*Acrocephalus scirpaceus*) chicks for 110 minutes, and measured begging rates every 10 minutes through this period. Both the number of begging calls given per time and the total gape area displayed by the brood increased steadily with deprivation time (figure 2.5).

ther experiments that have shown nestling begging to be influenced by either short-term deprivation or feeding include Von Haartman (1953) with pied flycatchers, Smith and Montgomerie (1991a) with American robins, Mondloch (1995) with pigeons, Leonard and Horn (1998) and Leonard et al. (2003a) with tree swallows (*Tachycineta bicolor*), and Maurer et al. (2003) with scrubwrens (*Sericornis frontalis*). The conclusion that begging reliably signals short-term need thus seems to be quite general.

Fewer studies have investigated the effects of long-term need on active begging. Working with nestling yellow-headed blackbirds, Price et al. (1996) assumed that long-term need would be greater in males than in females (because males grow to a larger final size) and would increase within either sex with declining condition (measured as the residuals of mass regressed on tarsus). Price et al. (1996) measured short-term need as the amount of food needed to satiate a chick to the point that it no longer opened its mouth to take food. Begging intensity was measured from the duration, number, and loudness of begging calls. Begging intensity was found to increase with increasing short-term need in the usual way, but the relationship between begging and long-term need was more complicated. Begging intensity increased as condition declined in females, but not in males. Controlling for hunger and condition, males begged more intensely than females on some measures (notably loudness of begging calls) but not others.

In a similar study, done with pigs rather than birds, Weary and Fraser (1995) used the residuals of mass regressed on age to measure long-term need, and manipulated short-term need by removing some piglets from their mothers' udders before milk ejection. When separated from their mothers, piglets low in mass for their age gave more calls than piglets high in mass. Piglets that were prevented from feeding also gave more calls than ones that had just been fed. Calling thus was affected by both long- and short-term need.

Iacovides and Evans (1998) manipulated both long- and short-term need in ring-billed gulls (*Larus delawarensis*). Young in a low long-term-need group were hand-fed to satiation four times a day from hatching on, while the young in a high long-term-need group were given only 75% as much food (by mass) at each feeding. Short-term need was manipulated by depriving young in both long-term-need groups for 1, 4, or 12 hours at each of five ages. Young in the high long-term-need group consistently gave more begging calls than did those in the low long-term-need group, and the discrepancy between the two groups increased with age, as would be expected if the long-term-deprived young were lagging further and further behind in condition. On the last day of measurement, 21–22 days post-hatching, the difference between the two long-term-need groups was extreme (figure 2.6). Within both long-term-need groups, begging call rates increased with length of short-term deprivation. Two other measures of begging, pecking rates and mean amplitude of begging calls, gave similar results, though the effect of long-term need on amplitude was not quite significant. Overall, this experiment shows very clearly that begging responds to both long- and short-term need in ring-billed gulls.

Wright et al. (2002) performed an experiment with a similar conceptual design using free-living pied flycatchers. Young were moved between nests to produce a set of large broods, with a mean of eight young each, and a set of small broods, with a mean of four young. Parents worked harder for the large broods, but still provided less food per chick. Chicks from the large broods

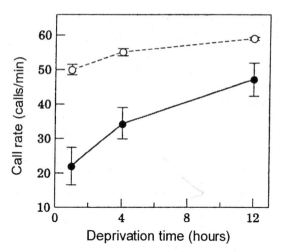

FIGURE 2.6. The response of begging call rate to short-term and long-term need in ring-billed gulls (from Iacovides and Evans 1998). Long-term need was manipulated by consistently feeding to satiation a high-condition group (closed circles), while limiting a low-condition group (open circles) to 75% of the first group's food intake. Short-term need was manipulated by varying deprivation time, that is, the time since last feeding. Begging call rate increases with deprivation time, showing the effect of short-term need; at any given deprivation time, begging call rate is greater in the low-condition group, showing the effect of long-term need.

thus were food-deprived over the long term, so that at any time they on average would have higher need, both short term and long term, than the young from small broods. Chicks from both sets of broods were then taken to the lab and fed to satiation, thus controlling short-term need. Subsequently, begging intensity increased faster in the chicks from the large broods, so that at 90 minutes after feeding they begged more intensely than the chicks from small broods, in terms of posture, call rate, and call volume.

Factors other than need have also been shown to affect levels of active begging. Kedar et al. (2000) trained nestling house sparrows (*Passer domesticus*) to beg at different levels by feeding one set of chicks as soon as they begged, even if they begged weakly, while feeding a second set only after they had begged at high intensity. Both sets of young were fed the same amounts of food, to control long-term need, and were tested for begging intensity under conditions that controlled short-term need. The chicks trained to beg at higher levels indeed begged more intensely, showing that learning can affect begging. Similar results were obtained by Rodríguez-Gironés et al. (2002) for magpies and great-spotted cuckoos (*Clamator glandarius*), that is, begging levels were again modified by past experience. These results can be interpreted as challenging the view that begging is a reliable signal of need (Kedar et al.

2000), and certainly they do show that there cannot always be a one-to-one correspondence between need and begging level. Nevertheless, showing that begging is not perfectly reliable does not justify concluding that it is not reliable at all, especially given that perfect reliability is a standard that no animal signal can be expected to attain. Instead, the abundant evidence that active begging reflects both short-term and long-term need justifies concluding that this signal is reliable in the sense of providing information that the receivers benefit from knowing.

The case of mouth color is more complicated, however. Parents have been shown to respond to mouth color in three species, canaries, barn swallows, and great tits, in all cases feeding young with red mouths preferentially. In canaries, Kilner (1997) found that the mouth color of chicks became significantly redder with increasing time since last feeding, and that the mouth color of chicks given supplemental food was significantly less red than that of control chicks whose food was not supplemented. Thus, mouth color in young canaries seems to be a reliable signal of short-term need, just as with active begging. By contrast, Saino et al. (2003) found that mouth color in barn swallow young did not respond to short-term food deprivation, but did respond to injection with sheep red blood cells, becoming less red relative to the mouth color of control young. Because treatment with sheep red blood cells mimics a mild infection, this experiment suggests that a red mouth in barn swallows is a signal of good health, rather than a signal of high need, as in canaries. In dark-eyed juncos (*Junco hyemalis*), mouth redness again did not change with food deprivation, but did respond to ambient temperature, becoming less red as temperature increased (Clotfelter et al. 2003). Note that parental response to gape color has not been tested experimentally in dark-eyed juncos. Finally, in great tits, the mouths of the young are orange-yellow, not red (Götmark and Ahlström 1997), and red in this species does not appear to be a signal of anything. Considering the four species together, it is impossible to form a consistent view of the reliability of mouth color, because we cannot form a consistent hypothesis on what information mouth color is supposed to convey.

Costs of Begging

The models of signaling between relatives that we have reviewed suggest that begging need not always be costly to be reliable, at least not in cases of partial signaling, where only limited information is transmitted using a small number of discrete signals (for example, hungry vs. not hungry). Many features of begging signals appear to be continuously graded, however, such as the rate of calling, the amplitude of calls, or the height to which a chick extends its head, suggesting that the partial-signaling assumptions may not apply to many real begging systems. Using the separation calls of piglets, which exhibit considerable variation in structure, Weary and Fraser (1995) explicitly searched for evidence of discretely different call types along four acoustic dimensions, including duration and three measures related to frequency characteristics. On only one measure, "peak frequency" (the frequency with maximum amplitude in the call), was there evidence of a multimodal distribution, with calls being especially common at two or three specific frequency ranges. Even for this feature, however, calls occurred in lower abundances at many different peak frequencies, and to maintain that there were only a few call types would thus require collapsing quite different calls into the same categories. Of course it is always possible that receivers perceive such graded signals in a categorical fashion (Harnad 1987), but for now there is no evidence that they do so with respect to begging signals.

If the assumptions of partial signaling are not met, then theory suggests that begging must be costly to be reliable. Empirically, two kinds of costs have been investigated: a direct, metabolic cost to the begging individual, and an indirect, survival cost through the attraction of predators.

McCarty (1996) was the first to measure the metabolic costs of active begging, using small numbers of subjects from each of seven passerine species. He placed one nestling at a time in a sealed chamber and measured its oxygen consumption during a 10–15-minute control period. Then during a second 10–15-minute period, he stimulated the nestling to beg by moving, tapping, and shading the chamber, while again measuring oxygen consumption. Oxygen consumed was converted to energy using the appropriate constant. Energy consumption was significantly greater during the begging period than during the resting (control) period, both in an analysis of the combined data from all seven species and in an analysis of the data from the one species with the largest sample size, the tree swallow. The proportional increase in energy consumption, however, was not impressive: in tree swallows, for example, the ratio of active metabolic rate (AMR) while begging to resting metabolic rate (RMR) was only 1.27.

Leech and Leonard (1996) independently measured metabolic costs of begging in tree swallow nestlings using a similar method. Their estimate of the metabolic scope (i.e., the ratio AMR/RMR) was 1.28, very close to McCarty's estimate. Metabolic scope increased with increasing begging intensity among older nestlings, but only weakly. Subsequently, Bachman and Chappell (1998) measured metabolic costs of begging in nestling house wrens (*Troglodytes aedon*), using a larger sample and a method (open-circuit respirometry) thought to be more accurate for measuring costs of brief behaviors such as begging. These authors arrived at an estimate of the metabolic scope for begging (AMR/RMR) almost identical to those previously given by McCarty (1996) and Leech and Leonard (1996) for tree swallows: 1.27.

The real impact of begging on energy budgets depends not only on the cost per time but also on the amount of time spent begging. Leech and Leonard (1996) estimated that the hungry tree swallow nestlings in their experiments spent on the order of 400 to 600 seconds per hour in begging, or about 11–17% of their time overall. The conditions of these experiments were artificial, however, in that the researchers continually stimulated begging and the nestlings were not fed; feeding of course tends to quiet begging. Bachman and Chappell (1998) estimated the time spent begging per day under field conditions, basing their estimates on videotaping of nestling house wrens within nestboxes. Here hunger levels should have been more natural, and the young must have been fed periodically by their parents. Nestlings were found to beg on average only 4–10 times per hour, depending on age, and begging duration averaged only 4–7 seconds. This generates an estimate of between 16 and 70 seconds spent begging per hour, or 0.4–2.0% of the overall time budget, an order of magnitude lower than the Leech and Leonard (1996) estimate. The two estimates are for different species, and we must expect time spent begging to vary not only between species but within species, with respect to environmental conditions such as food abundance. Nevertheless, on the basis of Bachman and Chappell's (1998) house wren data, our best guess is that nestlings usually spend very little time begging.

As a consequence of low energy costs per time combined with low amounts of time begging, the overall impact of begging on the daily energy budget is quite low. The best estimate is for house wrens, where begging appears to account for between 0.02% and 0.22% of the total energy budget, again depending on age (Bachman and Chappell 1998). Bachman and Chappell (1998) question whether costs this low are sufficient to maintain honesty. They calculate that if a nestling house wren were to dishonestly double its begging frequency (begs per hour), the extra cost would be more than compensated if the nestling was to receive just one extra feeding per day.

Another approach to assessing the metabolic cost of begging is to measure the effect of begging on growth. One advantage of this approach is that growth measurements typically integrate energy balance over a longer time period than can be used in measurements of metabolic rate; another advantage is that growth has a more direct and obvious relation to fitness than does energy expenditure. Kilner (2001) induced matched sets of canary young to beg for either 10 seconds or 60 seconds before each feeding over a 6-hour period. During the experimental period the chicks in the high-begging group lost more mass than the chicks in the low-begging group, but only when 8 days old; 6- and 10-day-old young showed no effect. Leonard et al. (2003b) performed a very similar experiment with nestling tree swallows, again inducing prolonged begging in an experimental group over a 6-hour period. The experimentals on average begged over three times as long (504 seconds) as the controls (152 seconds), but begging had no effect on mass gain. Rodríguez-Gironés et al. (2001b) induced prolonged begging over several days in groups of young ring doves (*Streptopelia risoria*) and European magpies. Increased begging had no effect on mass gain in the ring doves but lowered mass gain by 8% in magpies.

Finally, Kedar et al. (2000) trained one group of young house sparrows to beg more intensely than another group and found no difference between groups in mass gain. Overall, then, the evidence for an effect of begging on mass gain is positive in one species (magpies), ambiguous in a second species (canaries), and negative in three more (tree swallows, ring doves, and house sparrows).

Even though the energy cost of begging is low, it may be that begging nestlings are working as hard as they can—that is, that they are not capable of higher expenditures. Chappell and Bachman (1998) attempted to induce maximal activity in nestling house wrens by shaking and tilting the test chamber, causing the nestlings to struggle to maintain their balance and position. Nestlings appeared exhausted after short periods of these activity tests. A comparison of energy expenditures between begging and such periods of maximal activity was revealing: average energy-expenditure during begging was only half as high as during activity tests, but the maximal energy expenditure rates during begging closely approached the rates observed during activity. These results can be interpreted as showing that nestling house wrens are usually not expending energy at their maximal rate during begging, but that when they beg as intensely as they can they do approach their maximum level of effort.

A second potential cost of begging is that begging signals may attract predators. Presumably, such an effect would be due mainly to begging calls being overheard by a predator, although in open cup nests (as opposed to cavity or domed nests) a nestling that raises its head above the nest rim during begging might also attract visual notice. Haskell (1994) tested for an effect of begging calls on predator attraction in experiments in which he equipped artificial nests with miniature two-way radios from which calls could be broadcast. In one experiment, he set out 40 artificial nests on the ground and 50 in trees, the latter at a height of 2 meters. One quail egg was placed in each nest. From half the nests at each level he played begging calls of western bluebirds (*Sialia mexicana*), at a rate of 25 calls/min through the daylight hours. For the ground nests, playback had a significant effect on predation: after 5 days, 75% of nests with playback had been depredated, compared to only 25% of the silent control nests ($P = 0.011$). Playback had no effect on predation in tree nests: 48% of nests with playback were depredated compared to 63% of silent controls ($P = 0.20$). In a second experiment, this time with 50 ground nests, calls were played back at high rates (25 calls/min) from half the nests and at low rates (13 calls/min) from the other half. After 6 days of exposure, 64% of the nests with high call rates had been depredated, compared to 36% of nests with low call rates ($P = 0.022$). These results, then, support the idea that begging increases the chance of attracting predators, but only for ground nests, not for nests situated in trees.

Similar experiments have been performed by Leech and Leonard (1997) and Dearborn (1999). Leech and Leonard (1997) used tape recordings of begging tree swallows broadcast from artificial nests placed either on the ground or on

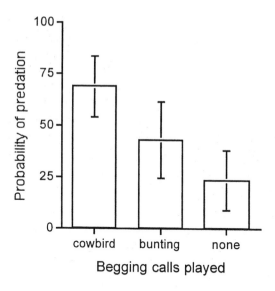

FIGURE 2.7. The risk of predation on artificial nests is influenced by the type of begging calls played at the nest (from Dearborn 1999). Nests were placed at sites appropriate for indigo buntings. At a third of the nests, a speaker played the begging calls of a brown-headed cowbird (a brood parasite), at another third a speaker played the begging calls of an indigo bunting, and the remainder served as silent controls. Calls were broadcast at natural rates and amplitudes: 300 calls/hour and 80 dB for the calls of brown-headed cowbirds, and 60 calls/hour and 74 dB for the calls of indigo buntings. The risk of predation increased with increasing rate and amplitude of the begging calls.

platforms about 1 meter off the ground. The nests with playback were depredated before paired silent control nests during 16 of 19 trials for ground nests (P < 0.01) and during 13 of 18 platform trials (P < 0.05). Dearborn (1999) placed 15 trios of artificial nests at heights of 0.2 to 0.5 meter in plants typically used for nesting by indigo buntings (*Passerina cyanea*). In each trio, speakers broadcast begging calls of the nest-parasitic brown-headed cowbird (*Molothrus ater*) at one nest and begging calls of indigo buntings at a second nest, while the third nest served as a silent control. Calls were played at amplitudes and rates realistic for the two source species, which meant that the cowbird calls were louder and more frequent than bunting calls. Predation rates were highest on the nests with cowbirds calls, intermediate on nests with bunting calls, and lowest on the silent controls (figure 2.7).

Existing evidence thus supports the idea that begging in nestling birds entails both energetic costs and costs associated with attracting predators, though a couple of important caveats are in order. First, the energy cost of begging admittedly is very low, as a proportion of the overall energy budget. Nonetheless, intensely begging nestlings seem to expend energy at a near maximal rate for their stage of development. The other factor keeping costs low is the low incidence of begging, which must be due in large part to the fact that receivers (i.e., parents) usually respond promptly to the signal. Nestlings may need to use a costly signal to demonstrate its honesty, but parents should have little interest in increasing the total cost of the signal to their offspring by delaying their response. Second, evidence to date supports the predation cost of begging only for nests built on or near the ground, but not for nests built higher in trees. The effect of begging on predators must depend on the hunting methods and sensory capabilities of the particular suite of predators operating wherever a particular species is nesting, and so should be expected to vary in magnitude with habitat and location as well as with nest height. That said, it is not intuitively obvious to us why the auditory and visual cues associated with begging are more prone to attract predators specializing on ground nests. What the predation experiments to date have demonstrated is that predation costs are possible, not that they always occur.

DECEIT IN BEGGING

The fact that in some cases begging has been shown to convey reliable information about offspring need does not eliminate the possibility that deceit also occurs in begging. To demonstrate deceit, one needs to show that the signal is in some instances unreliable in a direction that benefits the signaler. Demonstrating unreliability by demonstrating an imperfect correlation between begging intensity and need is not enough, because we expect such correlations always to be imperfect to some degree, if for no other reason than our inability to measure accurately either begging intensity or need. Fortunately, the occurrence of deceit in begging can be tested with a positive prediction, instead of the purely negative one of finding an imperfect correlation. This prediction derives from the circumstances under which offspring might be expected to benefit from exaggerating their own need, specifically when they are competing with others for their parents' attentions. If individuals increase their own begging in response to increases in the begging of others, this would be evidence that they are exaggerating their need, and therefore are in a sense deceiving the receivers.

To test for an effect of competitors on begging, Smith and Montgomerie (1991a) removed some but not all of the young from American robin broods, and deprived them of food for either 1 or 3 hours. After the food-deprived young were returned to their nests, the change in time spent begging among the nondeprived, control young was positively correlated with the change in time begging among the deprived young. This result does not prove that the control young respond to the begging of the deprived young; instead, it may be that the parents give a greater proportion of food to the deprived young the more those young beg, thus depriving the control young during the measurement period and causing them to beg in proportion to their own hunger. Smith and Montgomerie (1991a) showed, however, that a correlation in begging levels between control and deprived young is apparent quite early in the measurement period (during the first 30 minutes), making this alternative explanation less likely.

In an experiment performed with yellow-headed blackbirds, Price et al. (1996) paired focal chicks in turn with a larger hungry chick, a larger sated chick, a smaller hungry chick, or a smaller sated chick. Focal chicks begged significantly longer when paired with a bigger chick than when paired with a smaller one, and they begged longer when paired with a hungry chick than when paired with a sated chick. This experiment was performed in the lab, where chicks were fed by the researchers rather than by adult birds, so the effect of one chick on the other could not have been indirect, through an effect on parental feeding; instead, there must have been a direct effect of one chick's behavior on the other's begging.

Similar experiments have failed to show an effect of competitors on begging intensity in other species, notably European starlings. Cotton et al. (1996) manipulated hunger levels in starling chicks by feeding them either one, two, or three food items per 30 minutes. Two manipulated chicks were then placed together with one unmanipulated target chick. Feeding level had a significant effect on the manipulated chicks' latency to begin begging and on the amount of time they spent begging. Feeding level in the manipulated chicks did not, however, affect begging levels in the unmanipulated target chicks. Other experiments with starlings have had similarly negative results: begging level in one starling chick does not seem to affect begging level in another (Kacelnik et al. 1995, Cotton et al. 1996).

Another way to test for effects of begging on other nestlings is to use playback. Muller and Smith (1978) played recordings of begging calls to captive zebra finch (*Taeniopygia guttata*) young, alternating 2 minutes of playback with 5 minutes of silence for a total of 21 minutes. The frequency of begging by the young was significantly higher during the intervals with playback than during the intervals without. This result again suggests that chicks respond to the begging of others, but an alternative explanation is that parents respond to the playback by bringing food to the nest, and nestlings respond to the parents' arrival rather than directly to playback. Muller and Smith (1978) report that the nestlings often began begging immediately at the onset of the first playback session, whereas parents stepped up their feeding rate only after a lag, during which they acquired food. Begging frequency was even higher during the second and third playback intervals than during the first, perhaps because the effect of increased parental visits was added to the direct effect of playback. A direct effect of playback on the begging of zebra finch young thus seems likely, but an experiment that controls the response of the parents is needed.

Conclusions

Evidence that parents respond to begging is quite convincing, as is evidence that begging intensity can reflect need, either long term or short term. In many cases, then, begging appears to be a reliable signaling system, in which the signalers give the receivers information that the receivers benefit from knowing, and the receivers respond in a way that benefits the signalers. At the same time, there is evidence that signalers are capable of exaggerating their begging in response to competition from others. In general, the available models of signaling need between relatives that we reviewed in the preceeding section adequately account for empirical findings.

The most troubling question that remains, in terms of the match between models and reality, is whether begging really is costly in a meaningful way. Models suggest that begging need not be costly if the system is one of partial signaling, meaning that incomplete information on need is communicated because some individuals of differing need send the same signal (Bergstrom and Lachmann 1998). From empirical studies of begging interactions, however, it seems unlikely that begging systems really are partial signaling systems, though this question deserves more investigation. If, as seems more likely, begging communicates all levels of need, then the signal is expected to be reliable only if it has substantial costs, and whether the costs thus far demonstrated are high enough to fit this description is debatable (Kilner and Johnstone 1997).

Begging has been shown to attract predators, but existing evidence suggests that this cost may not apply in all cases. Begging clearly also has some metabolic cost, but the magnitude of this cost is not impressive, especially in comparison with other signals such as calling in frogs (Wells and Taigen 1986). Thus it is hard to avoid the impression that the costs of begging are unexpectedly low. Moreover, begging appears to be reliable even in those systems in which the costs seem especially minimal, as in the case of the red mouth color of begging young canaries. Kilner (1997) showed that mouth color in young canaries is a reliable signal of short-term need, and that parents respond to this signal. It seems highly unlikely that predators would detect this rather cryptic signal, and if predators cannot detect the signal it cannot have a predation cost. The signal presumably is produced by shunting blood to the mouth (Kilner 1997), an action whose energy cost seems likely to be negligible (although it is conceivable that a thermoregulatory cost is incurred, given that shunting blood to superficial capillaries commonly is associated with heat loss in vertebrates). Nevertheless, this aspect of begging seems to be as reliable a signal of need as most.

The apparent mismatch between data and theory is hard to evaluate, because begging models do not make testable, quantitative predictions about the magnitude of costs needed to generate reliable signaling. For now, it is perhaps better to depend upon the simplest versions of theory to give us a guide to what may be going on. Remember that in Johnstone's (1997) graphical models of signaling systems (figure 1.1), signal reliability can be stable either because the relationship between signal costs and signal intensity differs between signalers or because the relationship between signal benefits and signal intensity differs. For begging, the latter possibility is the relevant one. Some cost of the signal is needed to stabilize the signaling system (Godfray 1995), but it may be that much of the heavy lifting in maintaining reliability is done by the relationship between need, signal intensity, and benefit, rather than by the relationship between need, signal intensity, and cost. We return to this point in chapter 6.

Alarms

Animals give a variety of signals when a potential predator appears. A common interpretation of such signals is that they function to warn conspecifics. In cases where this interpretation has been shown to be correct, the signals represent another instance of signaling between individuals with overlapping interests: the signaler benefits from warning the receivers, usually because some are kin, and the receivers presumably benefit from being warned. Signals elicited by predators can have other functions, however; most important, the signal may be directed to the predator and may function to invite or (more likely) to deter pursuit. If signaling is from prey to predator, then signaling is between individuals with interests that are diametrically opposed, rather than convergent. Our first task in this section therefore is to identify cases in which predator-elicited signals are given between individuals with convergent interests, so that we can further analyze these particular signals. We then apply our standard analysis to this subset of alarms, examining the response of receivers, signal reliability, costs of signaling, and evidence of deception.

FUNCTIONS OF ANTI-PREDATOR SIGNALS

Among the many signals that prey give in the presence of predators, only a few have been studied in any depth with respect to function. We will concentrate in particular on two classes of signals: the alarm calls given by primates, passerine birds, and sciurid rodents, and the stotting and white-flash patterns of ungulates. These should be termed "anti-predator signals" rather than "alarms" to avoid prejudging whether or not they are given to warn others.

A large number of potential functions have been suggested for anti-predator signals (Sherman 1977, Bildstein 1983, Klump and Shalter 1984, Caro 1986a, Caro et al. 1995); these putative functions can be divided into two principal classes, on the basis of whether the intended receivers are predators or conspecifics. If anti-predator signals are aimed at predators, they may serve to invite pursuit before the predator is close enough to catch the signaler (Smythe 1970, Caro 1986a), to deter pursuit by signaling that the predator has been detected (Woodland et al. 1980, Bildstein 1983), to signal that the prey is in good condition and therefore cannot be caught (Dawkins 1976, Zahavi and Zahavi 1997), or to startle or confuse the predator (Walther 1969). If anti-predator signals are aimed at conspecifics, they may function to warn genetic relatives (Maynard Smith 1965), to manipulate other prey to behave in a way that makes the signaler safer (Charnov and Krebs 1975), or to reduce the overall chance of a successful attack and thus to lengthen the time until the predator returns (Sherman 1977).

The first set of hypotheses, those assuming that the signal is directed to predators, seems the best explanation for certain anti-predator signals in cursorial mammals, such as stotting and tail flagging. A white-tailed deer (*Odocoileus virginianus*) tail-flags by raising its tail to expose a small white rump patch and a more conspicuous area of white on the underside of the tail itself (Hirth and McCullough 1977). Several indirect lines of evidence support the view that this signal is given to predators rather than to conspecifics. First, deer are no more likely to give the signal when grouped with other conspecifics than when alone (Bildstein 1983). Second, the white patches are confined to a side of the animal's body (the rear) that is more often visible to predators than to other deer in the circumstances in which the display is given, that is, with the deer beginning to flee (Bildstein 1983, Caro et al. 1995). Third, females, who associate with kin, are no more likely to tail-flag than are males, who do not group with genetic relatives (Hirth and McCullough 1977, Caro et al. 1995); this fact weighs in particular against the hypothesis that tail-flagging functions as a warning to kin. Given the data suggesting that conspecifics are not the target of tail-flagging in white-tailed deer, the most likely explanation for the signal is that it deters predator attack by revealing that the signaler has detected the predator. As Caro (1995) notes, however, little evidence exists that the behavior of the predator is actually affected by this signal.

Stotting in Thomson's gazelles (*Gazella thomsoni*) is another anti-predator signal that seems to function principally in communicating to predators rather than to fellow prey. In stotting, an animal such as a gazelle holds its legs stiff while springing upward, so that all four feet leave the ground simultaneously (Walther 1969, Caro 1986a; see figure 2.8). Unlike tail-flagging, stotting can be seen regardless of the signaler's orientation with respect to the observer, so the signaler's orientation does not reveal to whom the signal is directed. Stotting is just as likely to be performed by solitary gazelles as by those in groups (Caro 1986b), which is more consistent with communication to predators than with communication to conspecifics. Moreover, there is some correlational evidence that predators attend to stotting: wild dogs (*Lycaon pictus*) tend to

FIGURE 2.8. Stotting in a Thomson's gazelle (from Walther 1969). On the left is a typical stotting posture; the animal has bounced upward with its legs held stiffly downward. In the middle, the animal has jumped unusually high, and is paddling its hindlegs. On the right, the animal is extending its front legs for landing. (Reprinted with permission from F. R. Walther, 1969. Flight behaviour and avoidance of predators in Thomson's gazelle [*Gazella thomsoni* Guenther 1884]. *Behaviour* 34:184–221. Brill Academic Publishers.)

single out for pursuit those gazelles stotting at lower rates, and when they switch attention in mid-hunt from one gazelle to another the switch is usually to an animal stotting at a lower rate (FitzGibbon and Fanshawe 1988). Alternative explanations for these observations can be imagined, however; for example, stotting might be correlated with condition (Fitzgibbon and Fanshawe 1988, Caro 1995), and predators might discriminate against prey in good condition after estimating condition by some signal other than stotting performance. Conclusive evidence that predators attend to stotting would require experimental manipulation of the signal, and it is easy to surmise why this has not yet been accomplished—one imagines a mechanical gazelle leading predators across the savanna, stotting as it goes.

In contrast to stotting and tail-flagging, alarm calls generally are thought to be directed to conspecifics rather than to predators. Much of the best data on function concerns various terrestrial species in the squirrel family, Sciuridae: ground squirrels, prairie dogs, and marmots. The evidence again consists mainly of observations of who calls and under what circumstances, as illustrated by the work of Sherman (1977, 1985) on Belding's ground squirrels (*Spermophilus beldingi*).

In many sciurids, different alarm calls are given for terrestrial and aerial predators. In Belding's ground squirrels the alarm for terrestrial predators is a trill, and that for aerial predators is a whistle (Sherman 1985). Sherman (1977) found that adult females are more likely than are adult males to give the trill in response to terrestrial predators. Because females are more site-faithful than males, and thus are more likely to be in the proximity of genetic relatives, the sex bias in calling can be interpreted as evidence that calling has evolved through kin selection (Sherman 1977, 1985). This interpretation is strengthened by the fact that females living in colonies with descendants (daughters and granddaughters) are more likely to trill than those living in colonies with no known relatives. In addition, those females with living nondescendant relatives (e.g., nieces and grandnieces) but with no living descendants are more likely to trill than are females with no known living relatives.

In contrast to terrestrial-predator alarms, alarms for aerial predators do not seem to be affected by the presence of kin. Males are as likely as females to whistle in response to aerial predators, and among females the tendency to whistle is affected by the presence of neither descendant nor nondescendant kin (Sherman 1985). The presence of conspecifics, however, does affect tendency to whistle: females living near conspecifics are more likely to whistle in response to aerial predators than are isolated females (Sherman 1985). Squirrels are more likely to call the more vulnerable they are, for example the farther they are from a burrow, and yet callers are less likely to be taken by a raptor than are noncallers (Sherman 1985). Sherman (1985) concluded on the basis of these data that the intended recipients of the aerial-predator alarm are other Belding's ground squirrels, and that the function of the alarm is to manipulate their behavior so as to increase the safety of the alarmer. In response to a whistle, listeners rush for shelter, "thereby creating predator-confusing pandemonium and a group in which to hide" (Sherman 1985). Note that under this "manipulation" hypothesis, there is no reason to think that receivers are harmed by their reaction to the alarm; rather, receivers "use the information for their own benefit, but in doing so make it possible for the caller to benefit even more" (Charnov and Krebs 1975, p. 110). Note also that selection will favor giving the terrestrial alarm to such conspecifics as are at hand, because giving it makes the caller safer; conversely, the fact that aerial alarms are rarely given to nonrelatives implies that giving these calls is costly to the caller.

Alarms in many other sciurids seem to function in a way similar to the terrestrial alarm of Belding's grounds squirrels, that is, in warning kin. The evidence again consists of who alarms and under what circumstances. Females tend to alarm more than males in those species in which dispersal is male-biased, so that females are more likely to remain in proximity to kin, for example round-tailed ground squirrels (*Spermophilus tereticaudus*) (Dunford 1977) and thirteen-lined ground squirrels (*S. tridecemlineatus*) (Schwagmeyer 1980). In thirteen-lined ground squirrels and Sonoma chipmunks (*Eutamias sonomae*), alarming by adult females increases dramatically at the time of year when their young emerge above ground (Schwagmeyer 1980, Smith 1978). In some species, such as black-tailed prairie dogs (*Cynomys ludovicianus*), the proximity of nondescendant kin seems to influence tendencies to alarm (Hoogland 1983), whereas in species such as Gunnison's prairie dogs (*Cynomys gunnisoni*) and yellow-bellied marmots (*Marmota flaviventris*) only the presence of descendant kin seems to have any effect (Hoogland 1996, Blumstein et al. 1997). There has been some controversy over whether selection for aid to nondescendant kin has ever been important in the evolution of alarming in squirrels (Shields 1980, Sherman 1980, Hauber and Sherman 1998, Blumstein and Armitage 1998). The evidence for aid to nondescendant relatives seems convincing for a few species (Sherman 1977, Hoogland 1983), but the point hardly matters in the present context, since warning either descendant or nondescendant kin constitutes signaling to individuals with overlapping interests.

Anti-predator calls in birds also appear to function, at least in part, in warning kin (East 1981). The best evidence comes from work on Siberian jays (*Perisoreus infaustus*), which live in groups consisting of a breeding pair plus offspring from a previous year, unrelated immigrants, or both. Griesser and Ekman (2004) tested the response of the jays to a hawk model flown over their heads at feeders. In this situation, breeding females were more likely to alarm when in the company of an offspring than when accompanied by an unrelated individual, showing an effect of kinship on calling. Breeding males almost always called for non-kin as well as for kin, however, suggesting that calling has other functions besides warning kin.

In primates, anti-predator signals may function both to warn conspecifics and to deter predators. In the Tai forest of West Africa, for example, various monkey species give alarm calls in response to chimpanzees (*Pan troglodytes*) and leopards (*Panther pardus*) (Zuberbühler et al. 1999). The hunting styles of these two predators are quite different: chimpanzees rely on pursuit to catch their prey, whereas leopards rely on surprise. Zuberbühler et al. (1999) showed that the monkeys give many more calls in response to leopards than to chimpanzees, and that a radio-tagged leopard tended to give up its hunt and move away after the monkeys it was stalking had alarmed. The interpretation suggested by these authors is that the alarms function in part to warn relatives, but that this can be accomplished by just a few vocalizations, with a larger number of alarms being given only when alarming has the additional benefit of informing a predator that relies on surprise that its prey have detected it (Zuberbühler et al. 1999).

Although the evidence that alarm calls function in warning conspecifics is stronger in sciurid rodents than in other groups, the evidence is still indirect, even for sciurids. Conclusive evidence that alarming increases the alarmer's inclusive fitness by saving kin, for example, would require experimental manipulations of alarming that result in measurable effects on the survival of kin; we know of no such experiments in any group.

RECEIVER RESPONSE TO ALARMS

To interpret evidence on receiver response to an alarm call, we first need to consider what kinds of information might be conveyed by the call. In the simplest case, an alarm conveys only that the signaler has observed a predator. Often, however, alarms convey additional information, such as the class of predator observed or the urgency of the threat posted by the predator. When alrms carry such additional information, the most advantageous response to the signal may vary, according to what call is given.

Domestic chickens (*Gallus gallus domesticus*), for example, possess separate, acoustically distinct alarm calls for aerial versus terrestrial predators (Gyger et al. 1987). When Evans et al. (1993a) presented hens in the laboratory with terrestrial alarms, aerial alarms, and control stimuli (noise), the subjects were much more likely to run to cover in response to aerial alarms than in response to the other two stimuli. Hens also responded to aerial alarms by scanning upward, and to both aerial and terrestrial alarms by scanning the horizontal plane, significantly more often than for the control stimuli (figure 2.9). The two alarm types also elicited different postures: hens crouched in response to aerial alarms and stood unusually erect for terrestrial alarms. Hens thus respond to both call types, and the fact that the responses seem adaptive with respect to the class of predator associated with the call helps to confirm the interpretation that the calls function as alarms.

Vervet monkeys (*Cercopithecus aethiops*) have a more elaborate system of distinct alarm calls for different classes of predators. These calls include a "leopard alarm" given for large mammalian predators, an "eagle alarm" given for large raptors, and a "snake alarm" given for snakes. Struhsaker (1967) first suggested that vervets respond differently to each of these alarms, a suggestion that Seyfarth et al. (1980) confirmed experimentally by playing the various alarms to vervets in the absence of any predators. Vervets were especially likely to look up in response to eagle alarms, to look down in response to snake alarms, and to run into a tree (if initially on the ground) in response to leopard alarms. Each of these responses is arguably adaptive, in the sense of increasing the safety of the responding individual.

As we stated in the preceding section, many ground-dwelling sciurids have been described as producing two types of alarm calls, one for terrestrial predators and one for aerial (e.g., Balph and Balph 1966, Melchior 1971, Greene and Meagher 1998). An alternative interpretation is that these different alarm types communicate degree of risk or response urgency, with the apparent association with predator type arising from the fact that there is more risk and less time to react for aerial than for terrestrial predators (Robinson 1981, Owings and Hennessy 1984). Various playback experiments have demonstrated that sciurids respond to their alarm calls (e.g., Blumstein and Armitage 1997, Hare 1998). As an example, California ground squirrels (*Spermophilus beecheyi*) typically responded to playback of their aerial-predator alarm by running to their burrow or to a boulder and then freezing (Leger et al. 1979). Playback of either aerial or terrestrial predator alarms elicited significantly more vertical postures in listeners than did playback of control sounds (Leger and Owings

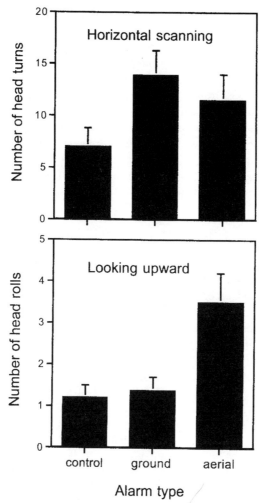

FIGURE 2.9. Visual monitoring by hens following playbacks of alarms (from Evans et al. 1993a). The two alarm types were ground and aerial alarm calls; the control was background noise. Both ground and aerial alarms caused significant increases in horizontal scanning relative to controls. Only aerial alarms caused a significant increase in looking upward.

1978). No statistically significant differences in response to terrestrial and aerial alarms were demonstrated (Leger and Owings 1978). These results demonstrate receiver response, but do not discriminate well between the response-urgency and predator-identification interpretations.

Suricates (*Suricatta suricatta*) give alarms that vary by both predator type and response urgency (Manser et al. 2001). These animals are social mongooses that inhabit semidesert areas in Africa. They possess different categories of alarms for terrestrial predators and aerial predators, and a third category—recruitment alarms—that they use for snakes and predator deposits such as urine or hair. Within each category, the acoustic structure of alarms changes in a consistent way as the level of urgency increases, that is, as the predator gets closer (Manser et al. 2001). Manser (2001) used playback experiments to show that suricates respond in qualitatively different ways to the different categories of alarms, for example grouping together in response to terrestrial alarms and approaching the speaker in response to recruitment alarms. At the same time, response strength varied with call urgency; for example, the time taken for the animals to relax increased with call urgency for both terrestrial and recruitment alarms. Receivers thus may respond to more than one kind of information encoded in an alarm.

Reliability of Alarms

With this background on the information that alarms convey, we can judge reliability of alarming, first, on whether a predator is actually present when the alarm is given, and, second, on whether the predator is of the correct class, in cases where the alarm refers to a specific class of predators, or whether the correct level of threat applies, in cases where the alarm conveys a specific level of response urgency. Gyger et al. (1987) provide data that can be used to apply these criteria to alarming by domestic chickens. These authors recorded alarms given by cockerels held in pens in an open field while simultaneously trying to observe any stimulus that might have elicited the alarm. For the 75 terrestrial alarms that they recorded (from five subjects), observers noted a possible stimulus for 63 (84%), whereas for 488 aerial alarms, possible stimuli were noted for only 268 (55%). These results imply that terrestrial alarms are more reliable than aerial alarms, but we cannot know how often the chickens actually alarmed for some stimulus that the human observers missed. Mistakes by the human observers were arguably more likely for aerial stimuli, which often were more distant and more fleeting than terrestrial stimuli (Gyger et al. 1987).

Of the 63 terrestrial alarms for which some stimulus was observed, 48 (76%) were associated with a terrestrial stimulus, 4 (6%) with an aerial stimulus, and 11 (17%) with a sound. Of the 268 aerial alarms for which a stimulus was observed, 239 (89%) were associated with an aerial stimulus, 11 (4%) with a terrestrial stimulus, and 18 (7%) with a sound. On this level, then, the two alarm types were both strongly associated with the "correct" class of stimuli. When we look at what these "correct" stimuli actually were, however, the chickens' performance becomes much less impressive, especially for aerial alarms. Of the 239 aerial stimuli associated with aerial alarms, less than 7% were hawks big enough to pose a threat to adult chickens, while another 33% were birds (such as crows and blue jays) that might take the eggs or young of

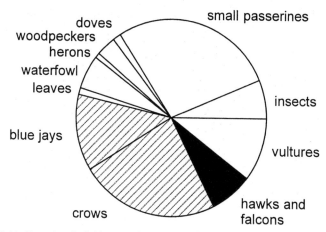

FIGURE 2.10. The stimuli eliciting aerial alarm calls from chickens held in an outdoor pen (data from Gyger et al. 1987). Black indicates species that might prey on adult chickens, hatching indicates species that might prey on eggs or chicks, and clear indicates stimuli that are unlikely to pose any kind of threat.

chickens (figure 2.10). Some of the remaining stimuli eliciting aerial alarms were ones that might understandably be mistaken for predators, for example vultures (9%), but others resemble predators in little or nothing beyond flight, for example small passerines (25%) and insects (6%). These naturally elicited alarms, then, were not terribly reliable, and we do not know whether the many instances of unreliability are best explained by some unknown benefit of false alarming, by mistakes of the human observers in locating the stimulus evoking the alarms, or by a combination of jumpiness and inaccuracy on the part of the alarmers.

Evans et al. (1993a,b) further investigated the stimuli eliciting alarms in chickens by presenting video images to male chickens under controlled laboratory conditions. Aerial predators were represented by animated drawings of hawk silhouettes shown on a video screen directly overhead, and terrestrial predators by a videotaped raccoon (*Procyon lotor*) shown on a video screen at ground level. Hawk animations elicited aerial alarms from 17 of 27 subjects and terrestrial alarms from none, whereas the raccoon elicited terrestrial alarms from 19 of 27 subjects and aerial alarms from none (Evans et al. 1993a). Hawk silhouettes elicited more calls the larger and faster moving they were (Evans et al. 1993b). Thus under better-controlled conditions, chicken alarms were more accurate in conveying the type of predator present and may also have contained information on the degree of threat.

Seyfarth and Cheney (1980, 1986) examined the reliability of alarms produced by vervet monkeys of different ages: infants (< 1 year), juveniles (1–5 years), and adults (> 5 years). All three age classes gave alarms almost exclu-

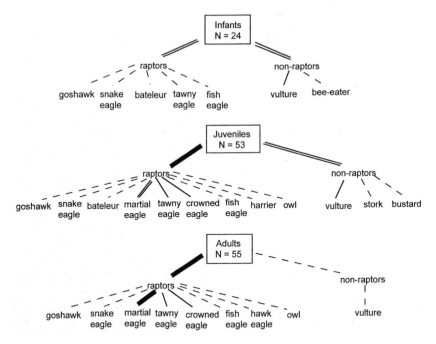

FIGURE 2.11. The stimuli eliciting aerial alarms from three age classes of vervet monkeys (from Seyfarth and Cheney 1986). The two confirmed predators are the martial eagle and the crowned eagle; the monkeys concentrate more of their alarms on these species as they advance in age. The thickness of the line indicates the number of alarms given to each stimulus: dashed line is < 5; single line is 6–10; double line is 11–15, thick line is > 15.

sively for animals of the correct general category: leopard alarms for terrestrial mammals, eagle alarms for birds, and snake alarms for snakes (Seyfarth and Cheney 1980). Within categories, however, incorrect signals were quite common, especially for younger animals. Adults gave 56% (31 of 55) of their eagle alarms for the two confirmed avian predators of vervet monkeys, martial eagles (*Polemaetus bellicosus*) and crowned eagles (*Stephanoaetus coronatus*), and most of the remainder to other raptors (figure 2.11) (Seyfarth and Cheney 1986). Infants and juveniles, by contrast, gave only 22% (17 of 77) of their eagle alarms for the two confirmed predators, and were considerably more likely than adults to alarm for nonraptors (figure 2.11). If we narrowly define correct alarms to include only those given to confirmed predators, then the eagle alarm overall is not impressively reliable; however, given the fact that most incorrect alarms are produced for stimuli that resemble true predators, and that the frequency of incorrect alarms decreases as age and experience increases, it seems likely that incorrect alarms represent simple mistakes rather than a signaling strategy that somehow benefits the signaler.

Age also affects the reliability of alarming in Belding's ground squirrels. Robinson (1981) examined production of alarm calls across several contexts, including social chases and encounters with terrestrial predators, aerial predators, and nonpredators. Combining the two types of alarms (terrestrial and aerial), adults gave 41% (26 of 63) of their alarms in one of the two predator contexts, whereas juveniles gave only 15% (11 of 73) of their alarms for predators. Adults were considerably less likely than juveniles to alarm for nonpredators, and also performed better than juveniles in associating the "correct" alarm type with the correct predator type. Nonetheless, despite the greater reliability of adult alarms, fully 50% of these were associated with no apparent cause.

Yellow-bellied marmots have been suggested to possess alarms specific to particular classes of predators, but Blumstein and Armitage (1997) found little evidence of this. The most common alarm (a whistle) was elicited by both aerial predators such as eagles and terrestrial predators such as foxes, as well as by nonpredators such as hares, pikas, vultures, and deer. This alarm was thus again relatively unreliable, even as to the existence of any threat, let alone the identity of the threat. The alarms were more reliable as regards the level of risk posed by the eliciting stimulus: two acoustic measures (starting minus ending frequency and number of whistles) successfully classified 80% of stimuli into high-risk and low-risk groups (Blumstein and Armitage 1997). A system that signals degree of risk, which may be common among sciurids, makes it more difficult for researchers to assess reliability, since we do not know a priori what risk categories the animals recognize, or where the boundaries between these categories are located.

Although the systems of alarm vocalizations we have reviewed all feature frequent false alarms, alarming nevertheless appears reliable enough to justify receiver response, in the sense that in all cases alarms are associated with real dangers on a substantial proportion of occasions. Escaping predation obviously has a huge fitness benefit, whereas the costs of responding to a false alarm in general seem to be low. Thus in signal-detection terminology, the cost of a missed detection is much greater than the cost of a false alarm, pushing the system toward a state of "adaptive gullibility" (Wiley 1994). The readiness of receivers to respond to alarms should leave these signaling systems vulnerable to cheating, if there is any benefit to deception, and if the costs of producing the signal are not too great.

Costs of Alarming

What costs might be associated with producing an alarm, whether reliable or not? Alarm calls are often short, single, and isolated, characteristics that should make them inexpensive to produce from an energetic standpoint. Any cost of alarms therefore seems likely to stem not from energy expenditure but from increased risk, specifically from increased risk of attracting a predator's attention and thereby being eaten.

In Belding's ground squirrels, observational evidence suggests that alarms do attract the attention of terrestrial predators. Sherman (1977, 1980) observed 22 instances in which a terrestrial predator stalked or chased a marked ground squirrel. Of squirrels observed to give the terrestrial alarm, 13% (14 of 107) were chased or stalked by the predator, compared to only 5% (8 of 168) of those that did not alarm. In a later analysis, Sherman (1985) found that 8% (12 of 153) of callers were actually captured by terrestrial predators, compared to only 4% (6 of 149) of noncallers. By contrast, aerial predators killed only 2% (1 of 42) of squirrels giving the aerial alarm, compared to 28% (11 of 39) of those not alarming. The data suggest that alarms given to terrestrial predators do put the alarmer at increased risk, whereas alarms given to aerial predators actually lower the alarmer's vulnerability. This difference is not due to the squirrels only giving aerial alarms when in a place of safety; to the contrary, squirrels were more likely to give aerial alarms the farther they were from a burrow and the closer they were to the raptor. The difference may be due in part to aerial predators being more susceptible to a confusion effect, and in part to the greater detectability and localizability of the terrestrial as opposed to the aerial alarm. The difference in risk costs associated with terrestrial and aerial alarms is also reflected in the social contexts in which they are given and their inferred function, as we discussed earlier. Terrestrial alarms are given disproportionately to relatives, presumably because giving these alarms is of net benefit to the signaler only when the signal's cost to self is counterbalanced by benefits to kin. Aerial alarms, with no cost to self, can be given with equal likelihood to related and nonrelated individuals.

Marler (1955) pointed out that the alarm calls given by small birds for aerial predators tend to have characteristics that make them hard to localize: they begin and end gradually (which denies information on location from differences in time of arrival at the predator's two ears) and they are of high frequency (which denies information from phase differences at the two ears). Alarm calls may also have properties that minimize their detectability by predators. Klump et al. (1986) measured the audiogram of the great tit and of its principal avian predator, the European sparrowhawk (*Accipiter nisus*). Although the predator's hearing was better than the prey's at low frequencies, the tit's hearing was 15–30 dB better than the predator's at the frequency of the great tit's aerial alarm. Consequently, the sparrowhawk probably is unable to hear the great tit's alarm at a distance greater than 10 m, whereas the great tit can hear another great tit's alarm at up to 40 m (Klump and Shalter 1984).

The acoustic properties of alarm calls may minimize the risk of alarming, but Sherman's (1980) data on terrestrial alarms in Belding's ground squirrels indicate that risk remains in at least some cases. This risk is a potential cost of alarming, but it is a cost that seems unlikely to enforce reliability, because the cost is much more apt to apply when the signal is honest (a predator has been observed) than when it is dishonest (no predator has been observed). An alarm given in the absence of an observed predator might still have a cost if an unobserved predator happened to be within hearing range of the signal, but the probability of this occurring must usually be low enough that the average cost would be negligible.

DECEIT IN ALARMING

Our discussion to this point has illustrated that receivers generally respond to alarms, even in the face of low reliability, and that producing alarms entails negligible costs for dishonest signalers; these are properties that would seem to invite exploitation through deception. All that needs to be added is a mechanism by which a signaler can benefit from eliciting an alarm response from receivers in the absence of a predator.

One way in which a signaler can benefit from a false alarm is for the alarm to move receivers away from some resource, thus allowing the signaler access. This benefit may apply to false alarming in the great tits (Matsuoka 1980, Møller 1988a). Great tits give alarms in the absence of predators when other birds, either of the same or different species, are feeding at a concentrated food source. As the other birds rush to cover, the alarmer flies in and takes food before the others return. Møller (1988a) found that dominant birds give false alarms to other dominant conspecifics, but not to subdominants, whom they are able to displace anyway. By contrast, subdominants give false alarms to both dominants and subdominants. In playback experiments, receivers responded equally to true and false alarms, suggesting that they do not discriminate between the two.

Munn (1986) found a similar pattern of false alarms in two tropical birds, the white-winged shrike-tanager (*Lanio versicolor*) and the bluish-slate antshrike (*Thamnomanes schistogynus*), though in these cases the dupes seem to be almost entirely heterospecifics rather than a mix of conspecifics and heterospecifics as with great tits. The shrike-tanager and the antshrike both commonly act as sentinels in mixed-species flocks, and are often the first to give alarms for actual predators. Both species also produce alarms when an insect has been flushed by flock members, giving the alarmer a momentary advantage in pursuing the insect as other birds respond to the alarm. In playback experiments, other species reacted equally to true and false alarms of the antshrike, again suggesting that these stimuli are not discriminably different (Munn 1986).

As Munn (1986) points out, there should be some upper limit to the ratio of false alarms to true alarms above which it would cease to be advantageous for receivers to respond to either. This critical ratio is expected to be quite high, because of the disproportion between the cost of responding to a false alarm (loss of a food item) and the cost of failing to respond to a true alarm (death).

Empirically, Munn (1986) observed 1.17 false alarms for every true alarm in shrike-tanagers, and Møller (1988a) observed 1.70 false alarms for every true alarm in great tits. Both these observed ratios may be well below the critical ratio that would push the system into instability. If any system of alarm signals has been overexploited by false alarming in the past, all that we would observe today is the absence of alarm signals, something we are unlikely to remark upon.

Møller (1989a) has suggested another possible benefit of alarm calling, in disrupting extra-pair copulations. Møller found that male barn swallows breeding in colonies give alarm calls in the absence of a predator when they return to their nests and find their female gone. Other barn swallows react to the alarms by flying out of the colony, and in five instances this response was observed to terminate an extra-pair mating involving the alarmer's female. Møller showed that false alarms were given almost exclusively during the female's fertile period, rather than before or after. Solitarily breeding males normally did not give alarms when their females were absent, but did so if they had previously been shown a taxidermic mount of a male barn swallow near their nests. All these findings support the interpretation of the alarms as a mate-guarding tactic, which benefits the male by minimizing his losses to extra-pair paternity.

A similar benefit for false alarms has been suggested by Tamura (1995) for the Formosan squirrel (*Callosciurus erythraeus*). This squirrel gives a repetitive, barking alarm call for mammalian predators, especially feral cats. Male Formosan squirrels also give a very similar call when tending an estrus female after copulation. In both cases, the calls are given in long bouts, averaging 7.0 minutes for the alarm calls and 17.0 minutes for the postcopulatory calls. The two calls are not discriminably different along a number of frequency and temporal dimensions, and the response of listeners to playback of the two calls seems identical: listeners climb higher in trees and then remain immobile until after the calls stops (figure 2.12). Tamura (1995) proposed that males benefit from the postcopulatory call because it immobilizes rival males and thereby increases the time interval to the female's next copulation. Studies with thirteen-lined ground squirrels have shown that the proportion of young sired by the first male to copulate with a female increases with increasing interval until the next copulation (Schwagmeyer and Foltz 1990), and lengthening this interval may thus have a reproductive benefit to the first male.

The Formosan squirrel's postcopulatory call might be interpreted as an aggressive signal rather than a false alarm. Alarmlike vocalizations are given during agonistic encounters with conspecifics in other sciurids, such as Belding's and California ground squirrels. In California ground squirrels, however, alarms given to conspecifics during chases differ acoustically from those given in response to predators, so that one class of alarms can be discriminated from the other (Leger et al. 1980). In Belding's ground squirrels, alarms are given in social contexts mainly by juveniles, and these alarms seem to elicit much

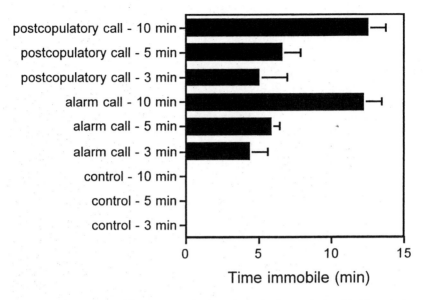

FIGURE 2.12. Responses of Formosan squirrels to playback of control stimuli (white noise), alarm calls, and postcopulatory calls (from Tamura 1995). The response measure is time spent immobile (mean + sd). The squirrels responded more strongly to both call types than to the control stimuli, the duration of response increased with playback duration for both calls, and no differences were observed in the responses to the postcopulatory calls versus responses to the alarm calls.

lower response than either alarms given to predators or alarms given for unknown causes (Robinson 1981), suggesting again that these alarms can be discriminated. By contrast, the postcopulatory, social alarms of Formosan squirrels do not differ acoustically from true alarms, and the two classes of calls elicit identical responses.

A second problem in interpreting postcopulatory calls as deceptive alarm signals is that their context is so specific: they are given only when multiple males have gathered on the territory of an estrus female and one male has copulated. If receivers cannot discriminate postcopulatory calls from real alarms acoustically, they could still evolve (or learn) to ignore all alarms heard within the postmating context. Presumably, the ratio of false alarms to true alarms within the postmating context is much higher than is the same ratio averaged over all contexts, making it more likely that a strategy of ignoring all alarms heard in this one context would be of net advantage. Existing evidence indicates that receivers do not follow this strategy, however (Tamura 1995). Apparently, encounters with predators are common enough, and the benefit of responding to alarms during encounters great enough, to make response to all alarms advantageous even in the postmating context.

Several possible cases of deceptive alarms thus have been found, in each of which an animal gives an alarm in the absence of any predator and benefits in one way or another from the responses of listeners. Note, however, that even though systems are known to exist in which animals give different alarms for different types of predators, no one has yet found deception through mislabeling of the type of predator. Presumably, the effect of this kind of deception would be to increase the risk experienced by listening conspecifics; for example, listeners might be made more vulnerable by being induced to respond as if a terrestrial predator were present, when in fact an aerial predator was present. The alarmer would only benefit in an evolutionary sense if this type of deceit was targeted exclusively to a narrow set of specific listeners, that is, a set that excludes genetic relatives and potential mates and includes only rivals. Such circumstances may be rare enough that this kind of system has never evolved.

Conclusions

Although the reliability of alarm calls in many systems appears to be low, alarms in general seem to be reliable enough to explain why receivers respond to them, especially given the huge potential cost of ignoring an accurate alarm. Most instances of unreliability probably are better explained as mistakes rather than as deceit—that is, a chicken or ground squirrel giving a false alarm usually receives no benefit from any response the alarm elicits. The only documented cost of alarming is that some types of alarm attract the attention of predators, increasing the chances that the alarmer will be captured and killed. This cost is much more likely to apply to honest alarms than to false ones, leaving little or no cost for false alarming. A signaling system in which false signals have little cost seems to invite deception, and a few candidate cases of deceit through false alarms have been identified.

Food Calls

Food calls are vocalizations given by an individual who has discovered a food source, and which typically have the effect of attracting other individuals to the food. In some cases the signaler and receiver are related individuals, as for example when mothers call to their offspring in gallinaceous birds (Collias and Collias 1967, Williams et al. 1968, Williams 1969). Here, it is obvious that the signaler's interests overlap those of the receivers, and therefore such examples fit easily into the conceptual framework of the present chapter. Slightly more problematic are cases in which the signaler calls to an entire group consisting of a mix of related and unrelated individuals, as occurs in many primates as well as some flocking birds. Callers in these cases may benefit from helping kin (Hauser and Marler 1993a), but other benefits may be at least as important if not more so. One possibility is that the caller benefits through increased safety from predation if he can gather group members around him; this may be the primary benefit of food calling in small, vulnerable species such as the house sparrow (Elgar 1986). Another possibility is that the signaler giving food calls benefits from attracting particular individuals that will form a coalition to help compete for the resource. Such coalitions are known to form among nonresident common ravens (*Corvus corax*) attracted to a carcass in winter, and usually without a coalition the nonresidents will be prevented from feeding by the resident pair (Heinrich 1988). Similarly, female bonobos (*Pan paniscus*) that are attracted to a food source by calling form coalitions with one another that allow them to compete successfully with males (van Krunkelsven et al. 1996). With both of these latter two functions, group safety and coalition formation, the interests of the signaler and the receiver can again be considered to be convergent, at least to some degree.

A rather different function applies when food calls are used by males to attract females for mating. This function can coexist with other functions within the same species. In California quail (*Callipepla californica*) and northern bobwhites (*Colinus virginianus*), for example, adults of both sexes give food calls during the parental period to attract their offspring to food, and during winter to bring group members together. Only adult males, however, give food calls during courtship to attract the opposite sex, with copulation often resulting (Williams et al. 1968, Williams 1969). In bonobos, females apparently give food calls to attract coalition partners, while males give food calls to attract females, at times trading access to food for copulation (van Krunkelsven et al. 1996). Even though food calls given to attract potential mates might better be considered signals between individuals with divergent interests, we will consider such examples in this chapter, along with the other instances of food calling.

A final possibility is that signalers give food calls as a way of proclaiming ownership of the resource. In ravens, as we have mentioned, coalitions of nonresidents are able to gain access to carcasses in winter in situations where solitary nonresidents would be repelled by local residents (Heinrich 1988). Nonresidents that find a carcass give a loud food call, the "yell," which has been shown in playback experiments to attract other ravens (Heinrich 1988). These observations accord with the view that the selective benefit of yelling lies in attracting a coalition, but a closer look reveals conflicting evidence. Heinrich and Marzluff (1991) investigated yelling in a captive flock of immature ravens, which were individually marked so that dominance relations could be observed. Dominant ravens were more likely to yell when a food source was discovered than were subordinates. Dominants seemed to suppress yelling in others through aggression, and subdominants were more likely to yell when dominants were removed or sequestered. Dominants attacked others that approached the food as well as those that yelled. Individuals were more likely to yell the longer they had been deprived of food. The observation that dominants suppress yelling in others and the fact that they attack others approaching the resource conflict with the view that yelling functions as a recruitment signal. Instead, the primary function of yelling seems to be to proclaim dominance and rights to the food resource.

Another case in which food calls function to proclaim ownership is provided by the white-faced capuchin (*Cebus capucinus*). These monkeys give "huh" calls when they discover food, especially fruit. Boinski and Campbell (1996) found that capuchins were more likely to call when their nearest neighbor was within 10 m than when there was no other individual this close, and that huh calls were predictably followed by an increase in the nearest-neighbor distance. Gros-Louis (2004) showed that capuchins that gave the food call when they encountered food were less likely to be approached by others than those that did not call, and were less likely to receive aggression. Again, this evidence suggests that the capuchin's food call functions to secure possession of food, rather than to attract others to feed.

Thus, it appears that food calling may represent (1) signaling between individuals with convergent interests in cases where food calls function to attract kin or to increase group safety, (2) signaling between individuals with divergent interests when food calls function to attract potential mates, or (3) signaling between individuals with opposing interests if food calls function to proclaim ownership of the resource. We consider examples of the first two categories of food calls in the remainder of this section.

RECEIVER RESPONSE TO FOOD CALLS

Observations of natural interactions show that offspring respond to parental food calls by approaching, for example in red jungle fowl (*Gallus gallus*) (Stokes 1971), California quail (Williams 1969), and northern bobwhite (Williams et al. 1968). In domestic chickens, chicks approach and increase their rate of pecking after their mother gives a food call (Wauters and Richard-Yris 2002). In spider monkeys (*Ateles geoffroyi*), subgroups at a foraging site were joined on 17% of occasions when one or more group members gave a food call, compared to only 5% of occasions when no call was given (Chapman and Lefebrve 1990). Playback experiments also demonstrate that food calls elicit approach by receivers. Elgar (1986), for example, observed the number of house sparrows arriving at an empty feeder within 5 minutes after broadcast of a house sparrow food call, broadcast of a human whistle, or a silent period. One or more house sparrows responded on 70% of trials with the food call, compared to 20% of trials with the human whistle and 30% of those with no playback.

In chickens, both observation and experiment support the hypothesis that hens respond to the food calls of males. Marler et al. (1986a) revealed food to a series of cocks in an apparatus that prevented a nearby hen from seeing the stimulus. Hens approached on 71% of 275 tests in which the test male gave one or more food calls, compared to 15% of 81 tests in which no calls were given. Evans and Evans (1999) played male food calls, contact calls, and alarms to isolated hens, using a loudspeaker that was hidden in the testing apparatus. Somewhat surprisingly, females tended to move away from the playback, regardless of call type. But the subjects were significantly more likely to look downward and fixate on the substrate in response to a food call than for other call types; these responses are behaviors typically shown during feeding. We can conclude that food calls generally elicit a response on the part of receivers that is consistent with an interpretation that the call conveys information about the presence of a food resource.

RELIABILITY OF FOOD CALLS

Food calls tend to be highly reliable, first in accurately signaling the presence of food, but also in many cases in signaling the quality or quantity of the available food. Again, some of the best data are from work on chickens. Wauters et al. (1999) observed levels of food calling in female domestic chickens presented with a food trough containing mealworms and mash (a preferred food), whole wheat and starter diet (the normal food), wood shavings, or nothing. Mean number of food calls emitted by the hens varied significantly with food treatment (figure 2.13). More calls were given for both foods than for either nonfood condition, and many more calls were given for the preferred food than for the normal food. In a later experiment, Wauters and Richard-Yris (2003) varied the amount of food presented to hens while holding food quality constant; subjects gave more calls for large amounts of food than for small. Marler et al. (1986a) presented four foods of differing desirability to male chickens, each housed with a female. The four food types, ranked a priori in terms of increasing desirability, were nutshells, peanuts, peas, and mealworms. Mean rates of food calling by males differed significantly across the stimulus types, with a ranking that matched the presumed desirability of the different foods.

Elowson et al. (1991) measured preferences for a variety of food items in captive cotton-top tamarins (*Saguinus oedipus*) using choice tests, and also found a strong correlation between mean preferences and mean calling rate in response to those foods. Other animals for which food calling has proved to be generally reliable include chimpanzees (Hauser and Wrangham 1987, Hauser et al. 1993), red-bellied tamarins (*Saguinus labiatus*) (Caine et al. 1995), and house sparrows (Elgar 1986).

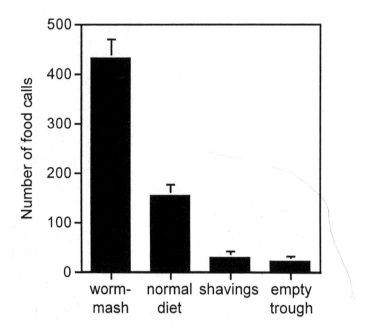

FIGURE 2.13. The number of food calls given by female chickens depends on the quality of the food present (from Wauters et al. 1999). These data (means + se) are from tests in which the hens could hear but not see their chicks. Worm-mash is a food preferred by the chickens over their normal diet of whole wheat and commercial starter diet. Wood shavings and an empty trough are conditions in which no food is present.

COSTS OF FOOD CALLING

Not much attention has been given to the costs of food calling. Some food calls are described as "soft," that is, of low amplitude, for example those given by gallinaceous birds (Williams et al. 1968, Williams 1969). Others, such as those given by red-bellied tamarins, are described as loud, but are also said to be short and given only one or a few at a time (Caine et al. 1995). Vocalizations with these properties are unlikely to be energetically expensive, nor are they likely to carry much risk of attracting predators. An interesting exception is the "yell" of common ravens, which is loud enough to be heard for several kilometers and can be given at rates of up to 40 per minute (Heinrich and Marzluff 1991). We argued above that the yell functions as an aggressive signal proclaiming ownership and dominance to individuals with opposing interests, rather than as a recruitment signal to individuals of converging interests, so it might be expected that a more costly signal is needed to ensure honesty. More work needs to be done before we can make definitive statements about the costs associated with food calling, but at present it seems fair to conclude that direct costs for the most part appear to be minimal.

Deceptive Use of Food Calls

Work on chickens has suggested that food calls may be used deceptively in situations where females respond by approaching food calls given by males. If approach on the part of the female increases the male's opportunity to mate, then there is an obvious payoff to the male for calling in the absence of food. Marler et al. (1986a) found that cocks often gave food calls in response to items that were not edible, specifically peanut shells. Furthermore, cocks were more likely to give food calls for these nonfood items if a female was present than in the absence of any receiver, and were even more likely to call for a strange female than for a familiar female (Marler et al. 1986b). These results provide strong evidence for deception: the signaler gives a signal that is appropriate if food is present even though food is not present, and the receiver responds in a way that (a) would be appropriate if food were present and (b) benefits the signaler. Why males give more deceptive calls to strange females than to familiar ones may be understood in terms of a type of model that we have not yet considered, one in which multiple interactions occur between signalers and receivers, allowing receivers to develop skepticism toward the signals of specific, unreliable signalers.

Individually Directed Skepticism

We introduce here a complication that we have previously ignored: the possibility that receivers can recognize the signals of individual signalers, remember the reliability of those individuals in past interactions, and adjust their present response in light of that past experience. If receivers refuse to respond in the future to the signals of individuals that have deceived them in the past, this behavior imposes a cost specifically on deceptive signals. Receivers might also retaliate directly, that is with aggression, against signalers they recognize as being consistently deceptive, imposing an additional cost on deceptive signaling. The costs we have considered previously have applied to all signals of a given form. For example, some signals have been considered to be costly because their form makes them energetically expensive or attractive to predators, or because receivers retaliate against any signal of that form. In contrast, the experience-based costs that we are now considering are ones that apply to a signal of a particular form only if that signal is paired with deception by a particular signaler. This property of experience-based costs ought to make them particularly effective in enforcing signal reliability.

Past experience might affect response to signals in mating and aggressive contexts as well as in signaling between relatives. We address this complication here because the model that best illustrates the effects of past experience builds on Maynard Smith's (1991a) Sir Philip Sidney game, described at the beginning of this chapter, and because some of the best empirical results on the effects of past experience concern signal categories covered in this chapter, namely alarms and food calls.

Maynard Smith (1991a) showed in the SPS model that cost-free reliable signals are possible as long as no conflict of interest exists between signaler and receiver. Another way of stating this result is that cheap, honest signals are possible as long as signaler and receiver rank the desirability of the possible outcomes of their interaction in the same order (Silk et al. 2000). Silk et al. (2000) argue that this last restriction can be relaxed if signaler and receiver are allowed to interact multiple times, and if the receiver remembers its past experience. Silk et al. (2000) base their model on a type of interaction in which a dominant primate approaches a subordinate and gives an affiliative call. The dominant may be peaceful, in which case calling is honest, or it may be intending to attack, in which case the honest course would be not to call. The subordinate can choose to stay and interact or to flee. The possible outcomes are a peaceful interaction, which is assumed to have a payoff of I to the signaler and i to the recipient; the recipient's fleeing, with payoffs D and d; and an attack by the dominant on the subordinate, with payoffs E and e. In an honest signaling system, the dominant signals only when peaceful and the receiver trusts the signal, staying if a signal is given and fleeing if it is not. In a one-time interaction, Silk and her colleagues find that honest signaling is possible only if both the signaler and the receiver rank the desirability of the possible outcomes in the same way, specifically with $I > D > E$ and $i > d > e$ (interaction is better than the receiver fleeing is better than an attack on the receiver). If multiple interactions are allowed, however, honest signaling is possible even if the interactants do not rank the outcomes in the same way. Suppose, for example, that $i > d > e$ still holds for the subordinate, but for the dominant $E > I > D$. In words, the best outcome for the dominant is to attack the subordinate, but a peaceful interaction is better than having the subordinate flee. In a one-time game, the dominant will want to deceive the receiver by giving the affiliative call and then attacking, receivers will then not believe the signal, and signaling will not be an ESS. In the multiple-time game, Silk and colleagues assume that the receiver will remember any instance of deception and will never again trust the signaler. A deceptive signaler will then reap a one-time advantage of $E - D$, but at a cost of $I - D$ for each of the subsequent interactions. If N, the number of future interactions, is large enough, then $N(I - D)$ will be greater than $E - D$. A one-time success at deception is therefore not advantageous over the long run, and honest signaling is favored.

The same kind of reasoning can be extended from affiliative calls to food calls or alarms. Suppose, for example, that a male benefits from giving a food call to a female because the female responds by approaching, thereby increasing the male's chance of mating. In a one-time interaction, a deceptive food call might well be favored. If the male and female interact multiple times, however, the male's one-time benefit from deceiving the female by calling in the absence of food might be outweighed by her subsequent reluctance to respond to his calls when food is present. This subsequent skepticism is a kind of cost imposed on the signaler, diminishing the temptation of the signaler to cheat, and stabilizing the signaling system. Similarly, a signaler might benefit from alarming in the absence of a predator by scaring others away from some resource, but that benefit could be outweighed if some of the receivers are its relatives and later fail to respond when the signaler gives alarms when predators are present.

It is possible that individual recognition may lead receivers to impose an even more direct cost on deceptive signalers, by aggressively retaliating against them. Consider the example of deceptive alarm signaling in great tits we discussed earlier (Matsuoka 1980, Møller 1988a). If dominant individuals recognized the alarm calls of individual subordinates, they might retaliate with aggression against those subordinates who have given an alarm when no predator is present. As far as is known, retaliation of this kind does not happen in great tits, presumably because alarms are not individually recognizable, and we know of no case in which this kind of receiver-dependent cost of deception has been documented. In theory, however, aggressive retaliation against deceptive individuals may represent another way in which individual recognition helps maintain the reliability of a signaling system.

For individual recognition to have an effect on signal reliability, it is necessary for receivers to learn and remember skepticism toward particular individuals. Hare and Atkins (2001) have demonstrated such individually directed skepticism for the alarm calls of Richardson's ground squirrels (*Spermophilus richardsonii*). Subjects were exposed to a series of ten playbacks of alarms recorded from two different individuals. The alarms of one individual were made reliable by consistently pairing them with the approach of a model predator, a stuffed American badger (*Taxidea taxus*). The alarms of the other individual were not presented with any predator, and thus ought to have appeared unreliable. The tenth playback in each series was a probe trial, in which neither stimulus was presented with a predator. On these probe trials, subjects showed significantly greater vigilance in response to the alarms of the reliable signaler than to the alarms of the unreliable one.

Cheney and Seyfarth (1988) demonstrated that skepticism toward individuals can carry over from one call type to a different but functionally similar call type in vervet monkeys. Their study focussed on "wrrs" and "chutters," calls used when one group contacts another. Vervet monkeys respond to both call types by orienting toward the caller and then looking in the direction that the caller is looking. Cheney and Seyfarth (1988) played a chutter recorded from a known individual to another individual from the same group, in order to determine a baseline response. On the next day, they played to the same subject a series of eight wrrs recorded from the same source individual. The wrrs were spaced at approximately 30-minute intervals. Subjects showed a diminishing response to these playbacks as the series continued, which is expected, since these calls were not paired with the appearance of another group. Cheney and Seyfarth then followed the series of eight wrrs with playback of a second chutter from the same source individual; here they found that the duration of response to the second chutter was significantly lower than to the baseline chutter. In a second experiment, the baseline and second chutter were recorded from a different source individual than the series of wrrs; here response to the second chutter was not lower than to the baseline, demonstrating that the reduced response to the second call type was individual-specific. Thus the monkeys learned to be skeptical of an individual's unreliable intergroup call, and the skepticism carried over from one intergroup call type to another, suggesting that an individual's signaling in general may come to be considered unreliable by members of its own social group.

The most compelling evidence to date that individually directed skepticism can work to maintain signal reliability comes from the work of Christopher and Linda Evans (2002, 2003) on food calls in domestic chickens. Evans and Evans played hens a series of food calls recorded from two different males. The calls of one male were made reliable by pairing them with food, while the calls of the other, unreliable male were presented without food. The hens quickly learned to discriminate, continuing to respond to the calls of the reliable male while ceasing to respond to calls of the unreliable male. Subjects remembered the discrimination for at least one day and generalized to other calls of the same individuals. The hens proved to have a threshold for detection of unreliability of 40%; that is, they discriminated against males who gave food calls without food more than 40% of the time. This value matches well with the observed occurrence of unreliable food calls in spontaneous interactions. We find this example especially compelling because the receivers show skepticism in a context where the signaler has a strong temptation to cheat, and receiver skepticism provides an explanation that is otherwise lacking for why the signal remains partially reliable.

In a second experiment, Evans and Evans (2003) found that hens failed to learn skepticism toward aerial alarms presented in the absence of a predator. Similarly, Cheney and Seyfarth (1988) showed that vervet monkeys fail to extend skepticism from an individual's eagle alarm call to the same individual's leopard alarm call, in the way that they do extend skepticism from one intergroup contact call to another. The common theme here is that animals are less likely to develop skepticism toward alarm calls than toward other categories of vocalizations, presumably because of the potentially disastrous costs of failing to respond to an honest alarm.

Conclusions

Receivers respond to all three categories of signals we have discussed in this chapter—begging, alarms, and food calls. All three categories are sufficiently reliable, in general, to explain this response—that is, the signals are accurate often enough to explain why it is advantageous for receivers to respond to them. Data on costs of signaling are less clear. Only for begging have costs been shown that might be high enough to ensure reliability; alarms and food calls seem by contrast to be relatively cost-free. Even for begging, however, signal costs seem remarkably low in most instances.

Models of signaling between relatives, reviewed at the start of the chapter, were formulated with begging in mind, and understandably apply more directly to this category of signals than to alarms and food calls. In a qualitative sense, the models are successful in explaining the mix of reliability and deceit seen in begging, with begging reflecting need, but capable of exaggeration when a begging nestling competes with other potential recipients of the parents' largesse. Again, the most troubling mismatch between theory and fact is the lower than expected costs of begging. The low cost of most begging suggests that the basic explanation for reliability of this signal type has more to do with the relationship between the signaler's need and the benefit it obtains from a receiver's response than with a relationship between need and signal cost.

To extend this idea informally to alarm signals, we note that models predict that signals between relatives can be reliable without cost if no conflict of interest exists. In the case of alarms, no conflict exists if both interactants benefit from a response when the signal is honest (a predator is present) and neither benefits when the signal is dishonest (no predator is present). This arguably is the case much of the time when alarms are given to warn relatives. Because no cost exists to enforce honesty, however, deception can be favored when a conflict of interest does appear—that is, when there is some benefit to eliciting receiver response in the absence of a predator, especially if the signal is directed at least in part to unrelated individuals.

We have given two rather different explanations for reliability arising from the first two signal types considered—a relationship between an attribute of the signaler (need) and signal benefit in the case of begging, and the lack of any conflict of interest in the case of alarms. A third explanation seems likely to apply to food calls, the final category of signals discussed in this chapter: that the occurrence of multiple signal exchanges makes deception on any one exchange disadvantageous because of the skepticism it will engender on subsequent exchanges. Thus we already have seen that multiple mechanisms must be invoked to explain signal reliability in animal communication, rather than just the handicap mechanism alone.

3

Signaling When Interests Diverge

This chapter concerns mating signals, those signals used by individuals of one sex to attract individuals of the other sex with the goal of inducing them to mate. In most mating systems, one sex does the bulk of the signaling, or "advertisement," while the other sex exercises choice among the signalers. A large literature exists that seeks, with considerable success, to explain why it is usually males that signal and females that choose (Bateman 1948, Trivers 1972, Clutton-Brock and Vincent 1991). Exceptions occur, but we shall take the usual pattern as a given, and as a shorthand we will speak in terms of males as the signalers and females as the receivers. The signaling models that we discuss ought in general to apply equally well to cases in which the usual pattern is reversed, so that females signal and males choose.

In the most general case, females are the more discriminating sex and are interested in mating with the best available male, whereas males are less discriminating and are interested in mating with any female. Male and female interests, then, are identical only in the singular case of the one male who is the best available; a female benefits from choosing him and he benefits from being chosen. For males of lesser quality, the interests of the sexes cease to be identical, in the sense that these lesser males would benefit from being chosen by females who would do better to choose someone else. It is in this sense that the interests of signaler and receiver diverge. Because of this divergence in interests, most signalers would benefit from exaggerating their quality, and questions of reliability and deceit become germane.

The literature on mating signals and the information these signals are thought to encode is voluminous. As in our treatment of signals where evolutionary interests overlap, we will confine our discussion here to a few kinds of signals that have been studied in considerable depth and that illustrate problems of reliability and deceit: carotenoid-based coloration, male bird song, and elongated tail plumage in birds. We begin with a review of the relevant theory.

Mating Signals: Theory

Grafen (1990b) offers a general argument for the stability of honest signaling of mate quality. Suppose that males possess some quality, q, that is important to females, but which females cannot observe directly. In addition, males possess an advertisement trait, a, which females can observe, and which can be made to reflect q. A male strategy consists of a function, $A(q) = a$, that translates a given quality into a specific level of advertising. Females have a complementary strategy, $P(a) = p$, by which they translate an observed level of advertisement to a perceived level of quality (p). Grafen (1990b) assumes that male fitness increases with p (i.e., that males benefit from being rated highly by females) and that female fitness increases as the difference between p and q decreases (i.e., that females benefit from getting their estimate of male quality right). Grafen then proceeds to show that an evolutionarily stable signaling system exists, given two conditions. First, the advertisement trait must have a cost, in the sense that male fitness decreases with increasing advertisement, holding female perception and male quality constant. Second, it must pay males to advertise more as q increases. If the benefit of advertisement (in terms of female perception) is the same for high-quality and low-quality males, then the latter condition reduces to the requirement that the marginal cost of advertising is higher for low-quality males than for high-quality ones. Grafen (1990b) thus shows that, given certain conditions on signal costs, a stable signaling system is possible in which females base their assessment of males on a male advertisement that reliably signals male quality.

Grafen (1990b) then turns the argument around, arguing that the existence of a stable system of mating signals implies that the signals must be both honest and costly. The argument is mathematical, but Grafen considerately provides a verbal description of the logic, as follows. At equilibrium, mating signals must be reliable, otherwise receivers would not use them, and signalers would not give them. If a signaler's level of signaling is at evolutionary equilibrium, then it cannot be of net benefit to him to increase his level of advertisement. Therefore, signaling at a higher level must be costly, and this cost must more than balance the gain (from impressing females) that the signaler would get from signaling more. We have assumed that signals are reliable (in the first step of the inverted argument), which means that a low-quality individual signals at a lower level than does a better individual. If signaling equilibrates at a lower level for the poorer signaler, then the cost of signaling must increase more steeply with increasing signaling level for the poorer individual than for the better one.

To test whether mating signals actually are reliable indicators of male quality, we have to be more specific about what we mean by quality. Many standards exist by which males could be assessed, and a given mating signal obviously could be highly reliable with respect to one standard while being completely unreliable with respect to others. In order to judge reliability, then, we need to know what standard females are applying, and this in turn requires that we understand the evolution of female preferences.

Numerous mechanisms have been proposed by which female preferences might evolve (Andersson 1994). In categorizing these, an initial distinction can be made between mechanisms requiring direct selection versus indirect selection (Kirkpatrick and Ryan 1991). Under direct selection, a female preference evolves because it affects the female's own survival or fecundity. Included here are female preferences for material benefits that a male might provide, such as nuptial gifts of food, access to a nesting or feeding territory, or provisioning of offspring. Under indirect selection, a female preference affects, not the female's own fitness, but the fitness of her offspring. Such indirect effects can occur through the acquisition of "good genes," that is, genes that increase the viability of the female's offspring. Indirect effects can also occur by the female's acquiring genes for traits that increase the attractiveness of her sons; preferences for these traits evolve through the "Fisher" or "runaway" mechanism. A third category of explanations for female preferences, apart from both direct and indirect mechanisms, suggests that preferences may be the result of sensory or neural biases of receivers that evolved outside the context of mate choice (Endler and Basolo 1998, Ryan 1998).

This litany is relevant here because only a subset of these mechanisms leads to female preferences for male signals that reflect any standard that is interpretable as male quality. A case to the contrary is provided by the Fisher mechanism. This mechanism was originally sketched out verbally by Fisher (1930), and has since been a favorite of modelers (e.g., Lande 1981, Kirkpatrick 1982, Pomiankowski et al. 1991). In simple terms, the mechanism starts with some female preference already existing at an appreciable frequency. Females having that preference mate with males having the preferred trait, setting up a genetic correlation between female preference and male trait in subsequent generations. Selection favors the preferred trait because of its advantage in mating, and the preference is favored indirectly because of its genetic correlation with the favored male trait. The preferred male trait may or may not be initially adaptive outside of mating, but regardless, it can be exaggerated to the point where it becomes maladaptive (Maynard Smith 1991b). At the end point, then, we would find females responding to a male signal that no longer reflects (if it ever did) male genetic quality (except as it relates to mating), male phenotypic quality, quality of direct, material benefits, or any other attribute that we could interpret as being related to "quality" in a functionally meaningful sense of that word. In such a case, it is pointless to inquire into the reliability of the signal.

It is an open question whether mating signals that have evolved via the Fisher mechanism exist at all, let alone whether they are common. Some theoreticians regard the logic of the mechanism as well established (Pomiankowski et al. 1991, Day 2000). By contrast, Grafen (1990a) comments that the idea "is too clever by half" and suggests that accepting the mechanism "without abundant proof is methodologically wicked." A candidate empirical case of Fisherian mate choice is provided by the sand fly *Lutzomyia longipalpis*, in which males provide no direct benefits to females, and male attractiveness does not correlate with female survival, female fecundity, or offspring survival, but does correlate with attractiveness of sons (Jones et al. 1998). Female sand flies thus benefit from their preference only through the enhanced mating success of sons, as predicted by the Fisher mechanism.

Receiver-bias mechanisms also produce female preferences for male traits that are not expected to reflect male quality. A variety of mechanisms have been named, differing in part on whether they emphasize peripheral sensory biases (Ryan et al. 1990), higher-level neural biases (Guilford and Dawkins 1991), or both (Endler and Basolo 1998). The biases in question may result from selection in contexts other than mating, for example selection for detection of prey, or they may be incidental, nonselected consequences of the ways in which sensory systems and brains are put together (Endler and Basolo 1998). All these receiver-bias hypotheses posit that the female bias evolves first, with the preferred male trait evolving later as a consequence of how the bias affects mating. At least initially, then, there is no reason to expect the preferred male trait to be correlated with any aspect of male quality; such a correlation certainly is not needed to make the mechanism work. Some good candidate cases of receiver bias have been proposed in which there is evidence that a preference antedates the preferred trait, involving for example preferences for call features in *Physalaemus* frogs (Ryan et al. 1990, Ryan and Rand 1993), for visual ornaments ("swords") in swordtail fish of the genus *Xiphophorus* (Basolo 1990, 1995), and for vibratory signals in the water mite *Neumania papillator* (Proctor 1991, 1992).

This leaves two categories of mechanisms that produce female preferences for traits that are expected to reflect male quality in some sense: direct selection favoring preferences for benefits that increase a female's survival and fecundity, and indirect selection favoring preferences for good genes. Grafen (1990a) provides a population-genetic model for direct selection on a female preference, based on the game theory model discussed above (Grafen 1990b). In the genetic model, male quality is environmentally determined, and quality thus has no genetic component that can be inherited by offspring. One haploid genetic locus determines both the function, $A(q)$, by which males translate quality into advertisement and the function, $D(a)$, by which females determine their probability of mating on the basis of male advertisement. Advertising is costly, and the cost is greater for low-quality males than for high-quality males. Because the number of offspring produced by a female increases with increasing male quality, direct selection favors female preferences for good males. Grafen seeks a pair of functions, $A^*(q)$ and $D^*(a)$, which are not invasible, in the sense that when these functions predominate in a population, no other rule can increase in frequency. Two pairs of functions are found. One is a nonsignaling equilibrium, at which all males signal at the lowest level possible and females ignore the signal. The other is an honest signaling system, in which the level of male signaling strictly increases with male quality, and the probability that a female will mate increases strictly with male advertising level. Grafen argues that the Fisher mechanism can play no role in the form of this equilibrium, because at equilibrium no genetic variability exists, and thus no genetic correlation between signal and preference is possible. Grafen's model makes narrow assumptions about the genetic system controlling male trait and female preference, but other models, assuming other systems of genetic control, have also found that female preferences can evolve via direct selection (Kirkpatrick 1985).

In good-genes models, a male trait signals "viability," by which is meant fitness exclusive of mating success; this exclusion is made to separate the good-genes mechanism from the Fisher mechanism. The logic of good-genes models runs as follows (Andersson 1994). Assume that a male trait exists that is reliably correlated with male viability, and that both this advertisement and viability have a genetic basis. If a genetically determined preference for the male advertisement appears in females, then females with the preference will mate with males having genes for both the advertisement and high viability. Consequently, a genetic correlation is produced between all three traits: male advertisement, female preference, and viability. Natural selection directly favors high viability, and the preference and the advertisement will be favored indirectly because of their genetic correlation with viability.

How well this model works depends on how the correlation between advertisement and viability arises and is maintained. According to the handicap principle, the advertisement must be costly to be reliable, but we can imagine different ways that a cost could be imposed. One is for the advertisement to have a survival cost, varying with male quality, and of a magnitude sufficient to have a significant effect on the chance that a low-quality male survives to mating. Survival to mating then demonstrates that a handicapped male has good genes for survival. An ornament that works this way, as a sort of survival test, has been termed a "pure epistatic handicap" (Maynard Smith 1991b) or "Zahavi's handicap" (Pomiankowski and Iwasa 1998). The latter name implies that this is how Zahavi imagined handicaps to work, which is a reasonable interpretation of his early papers. For example, Zahavi (1975, p. 211) states that "since good quality birds can take larger risks it is not surprising that sexual displays in many cases evolved to proclaim quality by showing the amount of risk the bird can take and still survive." Population-genetic models, however, generally suggest that female preferences for Zahavi's handicaps are unlikely to evolve (Maynard Smith 1976b, 1991b, Iwasa et al. 1991). One problem is that the link between signal expression and genetic quality is too indirect; a second problem is that the benefit a female receives by choosing a handicapped male, in terms of her offspring inheriting genes for high survival, is diluted by her sons also inheriting an ornament detrimental to survival.

Alternatively, a handicap might be "condition-dependent," that is, expressed only if the male has good genes for viability (Andersson 1986, 1994). Models indicate that female preferences are more likely to evolve for a condition-dependent handicap than for a Zahavi's handicap, in part because the link between the signal and genetic quality is more direct, and in part because the cost of the signal is not paid by all of a female's sons, but only by those that inherit genes for high viability as well as genes for the signal (Iwasa et al. 1991, Pomiankowski and Iwasa 1998). A third possibility is a "revealing handicap," which is expressed to some extent in all males having the genes for the handicap, but which is expressed to a perceptibly greater degree in those males that also have genes for high viability (Maynard Smith 1991b, Andersson 1994). Here the link between the signal and genetic quality is again direct, but all sons inheriting genes for the signal pay the cost of the signal. Genetic models are again favorable for the evolution of female preferences based on revealing handicaps (Maynard Smith 1985, Andersson 1994). The distinction between condition-dependent and revealing handicaps is substantive, but is difficult to maintain operationally, because of the difficulty of assessing independently whether males with the handicap have genes for high viability. We will lump the two categories together as condition-dependent handicaps or indicators.

In summary, if female preferences evolve as receiver biases or via the Fisher mechanism, then the preferred male traits cannot be expected to reflect male quality in any sense, and questions of reliability and deceit do not apply. If, on the other hand, female preferences evolve by direct selection or through a good-genes mechanism, then the preferred traits should reflect male quality, and questions of reliability and deceit become relevant. How then are we to distinguish between preferences evolved under one set of mechanisms versus those evolved under the other? Unfortunately, the primary criterion for distinguishing preferences evolved via the Fisher mechanism from those evolved via good genes and direct-selection mechanisms is that under the latter we expect the preferred traits to signal male quality and under the former we do not. Thus we are left in the sorry position of predicting that male mating signals will be reliably correlated with male quality except when they are not; theory covers us either way.

Another problem arises from the fact that male "quality" is likely to be multidimensional in many systems. Under a good-genes mechanism, a male signal should reflect male genotypic quality, in the sense of viability that can be inherited by offspring. Under direct selection, a preferred male trait might reflect the quality of male-controlled resources or of future paternal care. There is no inherent reason a single male display trait cannot signal both aspects of quality at the same time. The rate at which a male produces a vocal display, for example, may be correlated with his physiological condition and hence also with his viability, his future rate of offspring provisioning, and the density of food on his territory. Because a female assessing vocal rate may be interested in any one of these aspects of quality, we cannot be sure which correlation is most relevant to an analysis of reliability. We must keep these complications in mind when assessing the reliability of mating signals.

What about deceit in mating signals? Suppose we observe a system in which females are interested in male genetic quality, and in which males signal their quality via some advertisement. Suppose also that the cost of the signal increases as male quality decreases, so that the signal is on the whole reliable. Does room exist for deceit in such a system? Kokko (1997) examined this question for the case of an age-dependent advertisement, that is, a display character that increases with age. Her model assumes that there are only two classes of male quality. Males are given a condition that depends on their quality, and they can allocate part of that condition to the advertising trait, which is all that the females can assess. Male survival depends on that part of condition not allocated to advertising, so a tradeoff exists between advertising and survival. Some proportion of advertisement can be carried over to the next year, allowing advertisement to increase with age. A male's fecundity depends on his advertisement relative to the rest of the males in his population and on the strength of female preference for above-average advertisement. A male's strategy is defined by his allocation to advertising at each age. Given this setup, Kokko (1997) looked for strategies that are evolutionarily stable.

The outcome of Kokko's model depends on parameters such as the strength of female preference, the disparity in quality between the two male quality classes, and the proportion of the advertisement allowed to carry over between years (figure 3.1). Signaling systems are not necessarily maintained, for example in cases where female preferences are weak. Signaling systems, when they do exist, are not always strictly honest, in the sense that the advertisement levels of high-quality males are not always higher than those of low-quality males at every age. Nevertheless, the signaling systems are always "honest on average," meaning that dishonesty at one age is more than compensated by honesty at other ages, and a female thus increases her overall chances of obtaining a high-quality male by paying attention to the advertisement (Kokko 1997). The value of Kokko's (1997) model is in showing that some level of dishonesty about male genetic quality can result from adaptive signaling choices rather than just from sloppiness in the signaling system.

A second model by Kokko (1998) looks at the signaling end of the system only. This model suggests that dishonest signaling about genetic quality is particularly likely if there is some threshold, in the allocation to self-maintenance, below which the future fitness component of the signaler is near 0. In simpler terms, those males possessing few resources to spend on self-maintenance might as well spend their all on advertisement, because they are not going to survive anyway. In this way, males of very low quality can end up advertising at a higher level than males of somewhat higher quality. Kokko (1998) notes that dishonesty is explained here by that fact that this case does not meet the requirement that costs of signaling are greater for

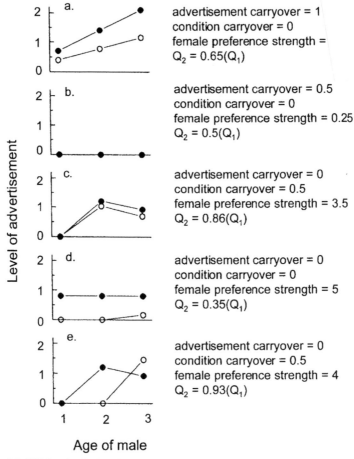

a. advertisement carryover = 1
condition carryover = 0
female preference strength =
$Q_2 = 0.65(Q_1)$

b. advertisement carryover = 0.5
condition carryover = 0
female preference strength = 0.25
$Q_2 = 0.5(Q_1)$

c. advertisement carryover = 0
condition carryover = 0.5
female preference strength = 3.5
$Q_2 = 0.86(Q_1)$

d. advertisement carryover = 0
condition carryover = 0
female preference strength = 5
$Q_2 = 0.35(Q_1)$

e. advertisement carryover = 0
condition carryover = 0.5
female preference strength = 4
$Q_2 = 0.93(Q_1)$

Level of advertisement

Age of male

FIGURE 3.1. ESS levels of advertisement at different ages of males of high quality (filled circles) and low quality (clear circles), from Kokko (1997). Males are given a certain condition each year, which they can invest in advertisement or survival. Base condition is determined by male quality and is either high (Q_1) or low (Q_2). Female preferences are based on advertising. The model allows males to carry over between years a proportion of their advertising (**a, b**) or of their condition (**c, e**). Versions of the model also differ in the strength of the female preference and in the degree of disparity between Q_1 and Q_2. Certain combinations (**a**) predict advertising levels that increase monotonically with age and that reliably reveal condition at every age. Other combinations, especially with weak female preferences (**b**), predict no advertisement. Advertisement can be reliable about quality at each age without always increasing with age (**c, d**). Finally, advertisement can be unreliable about quality at certain ages (**e**) while still being honest on average, in the sense that females choosing on advertisement obtain mates of higher mean quality than females ignoring advertisement, averaging over the age classes. (Reprinted with permission from H. Kokko 1997. Evolutionarily stable strategies of age-dependent sexual advertisment. *Behavioral Ecology and Sociobiology* 41:99–107. Springer-Verlag.)

males of lower quality. The overall signaling system still is expected to be "honest on average."

Theory then leads us to the expectation that male signals can be honest about aspects of quality that females are interested in, but also suggests that empirical evaluation of honesty will be made difficult by our own uncertainty about what it is that interests females. Theory also allows some level of deception in mating signals, but it is again the case that identifying occurrences of deception will be difficult if we are unsure about what the signal is "supposed" to convey, as is often the case in empirical studies.

Carotenoid Pigmentation

Carotenoids are a class of pigments defined by the common chemical structure of a linear chain of carbon atoms linked by alternating single and double bonds, with a six-carbon ring at each end (Hill 2002). These pigments absorb light in the blue and violet range of the electromagnetic spectrum, and therefore reflect or transmit light in the range of yellow, orange, and red. The color of a given carotenoid depends in part on the number of alternating double-bonded carbon pairs in the chain, deepening from yellow to orange to red as the number of such pairs increases (Fox 1979). When combined with proteins, carotenoids also can reflect or transmit purple, green, and blue wavelengths (Olson and Owens 1998).

Carotenoids are found widely in animals and, in addition to their role as pigments, are involved in a variety of physiological functions (Olson and Owens 1998, Hill 1999). Animals lack the biochemical pathways to manufacture carotenoids, so they must obtain them by eating the plants, bacteria, and fungi that can synthesize them de novo (Fox 1979). This fact, together with their physiological usefulness, may explain in part why carotenoid-based pigments are so often employed in visual mating signals.

Before discussing empirical work on the use of carotenoids in visual signaling, we need to acknowledge a major problem in interpreting functional studies of animal color patterns in general, including those produced by carotenoid pigmentation. The problem is that the color vision of other animals differs from that of humans, and consequently the perception of color patterns by other animals may be radically different from our own (Endler 1990, Bennett et al. 1994). Birds, for example, are sensitive to ultraviolet wavelengths that we cannot see at all, and have at least four cone types to our three (Bennett et al. 1994). Having an extra cone type may increase the dimensionality of color vision, producing "a qualitative change in the nature of color perception that probably cannot be translated into human experience" (Bennett et al. 1994, p. 851). Because of these and other differences in color perception, the use of human observers to assess the colors of other animals may be seriously misleading. In addition, the use of dyes, paints, or the like to manipulate color patterns may produce unintended changes in the pattern as perceived by the animals involved (Bennett et al. 1994). Unfortunately, much of the available empirical work relies on subjective ratings of color and uncalibrated experimental manipulation of color, raising concerns about how strongly this work can be interpreted.

The issue of the subjectivity of color perception can be avoided to some extent by assessing color patterns using spectral analyzers that quantify the wavelength of reflected light, and by manipulating colors using chemicals or filters that have been calibrated to block specific wavelengths in a selective fashion (Endler 1990, Cuthill et al. 2000). Although such methods have revolutionized the study of UV patterns in birds (Cuthill et al. 2000), they have not led to the widespread rejection of previous conclusions on carotenoid color patterns. This fortunately suggests that human ratings of carotenoid patterns and traditional methods of manipulating those patterns are sufficient for many purposes. The studies we report on have used subjective rating of color, unless otherwise noted below.

RECEIVER RESPONSE TO CAROTENOIDS

Evidence that carotenoids play a role in signaling has been found in many species, but we concentrate on three examples that have been especially well studied: one bird, the house finch (*Carpodacus mexicanus*), and two fishes, the three-spined stickleback (*Gasterosteus aculeatus*) and the guppy (*Poecilia reticulata*). Guppies are small fish familiar from aquariums and native to Trinidad and nearby areas of South America. Male guppies exhibit complex patterns of spots, varying among individuals and populations in size, position, and color (Endler 1980). Carotenoids are responsible for the coloration of orange and red spots, whereas spots of other colors contain different pigment types (Kodric-Brown 1985). Guppies have a mating system that includes multiple matings by females and relatively little male-aggressive competition, features that would seem to give ample scope for female choice.

Kodric-Brown (1985) and Houde (1987) both found that the mating preferences of female guppies correlate with naturally occurring variation in the orange coloration of males. In one experiment, Houde (1987) placed pairs consisting of one male and one female together in aquariums and measured time to mating. In all, 40 males were used, each categorized to one of four classes on the basis of relative extent of orange coloration. Females on average mated more quickly the more orange the male, and the effect was a strong one. In other experiments, females given choices between males were more likely to approach and associate with males the greater the extent of their orange coloration (Kodric-Brown 1985, Houde 1987).

Using natural variation in male coloration in such experiments leaves open the possibility that females judge males on some other trait that is correlated with extent of orange. This alternative can be eliminated by manipulating the focal trait experimentally. Kodric-Brown (1989), for example, used diet to manipulate orange coloration in guppies. Thirty-one pairs of sibling males were divided into two groups soon after birth. One group was given a base diet while the other was given the base diet plus two synthetic carotenoids. Six months later, males with the carotenoid-supplemented diet had red and yellow spots considerably brighter than those of the control males. Diet had no effect, however, on the area of spots. In trials where one female interacted with one experimental and one control male, females mated four times more often with experimentals than with controls. Grether (2000) also found that female guppies preferred males raised on high-carotenoid diets; however, orange chroma (measured with a spectroradiometer), rather than brightness, was the color measure that increased with increasing diet carotenoids.

Results of experiments in which diet is manipulated are consistent with an effect of orange coloration on female preferences, but one alternative remains—that carotenoid supplementation changes some phenotypic character other than coloration, and that females attend to this correlated character rather than to color. This alternative is addressed by experiments performed with our second species, the three-spined stickleback. Males in this species assume a red or orange-red ventral coloration during the breeding season. Semler (1971), working in a population polymorphic for this breeding coloration, gave females a choice between pairs of males, one red and one non-red, that had built their nests at the opposite ends of aquariums. Choice was recorded when a female laid her eggs in one of the nests. Female sticklebacks showed a more than two-to-one preference for the red males (figure 3.2). Semler (1971) repeated the experiment, using not naturally red males but males that had been painted red with lipstick or nail polish. Females showed as strong a preference for the artificially red males as they had for the naturally red ones (figure 3.2). Female sticklebacks thus must attend to the red coloration.

A similar conclusion was reached by Milinksi and Bakker (1990), using a much different design. Female sticklebacks were given a choice between pairs of males, each male in his own tank, and differing naturally in color. A female, in a third tank, showed her preference by assuming the head-up courtship posture while oriented toward one of the males. Females showed a preference for the redder of the two stimulus males when tested under white light. Females showed no color preference, however, when tested under green light, where the color difference could not be perceived. The design eliminates the possibility that females prefer red males on some cue other than color, since the preference disappears when they cannot perceive the color differences. Other work has shown that females prefer redder males in the field as well as in the lab, with redness measured objectively using a densitometer (Bakker and Mundwiler 1994).

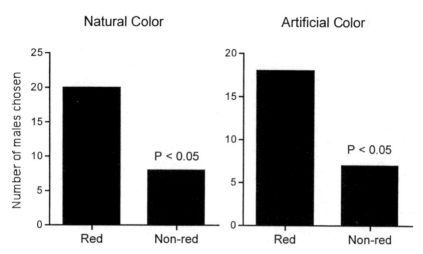

FIGURE 3.2. Female preference for red males in three-spined sticklebacks, from Semler (1971). Each female was given a choice between a red and a non-red male. The males had nests at opposite ends of an aquarium, and a choice was scored when the female laid eggs in one of the nests. Significantly more females chose the red male, both when the red color was natural and when it was artificially imposed. The strength of the preference was very nearly equal in the two conditions.

Hill (1990, 1991, 2002) has used a variety of approaches to demonstrate a female preference for red males in house finches. Male house finches have three patches of carotenoid pigmentation, one on the crown, one on the throat and breast, and one on the rump. Each can vary in color from pale yellow to bright red (Hill 1992). Hill (1990) demonstrated that captive female house finches preferred to associate with redder males when males varied naturally in color, when male color was manipulated through their diet, and when color was manipulated with dyes. Among free-living house finches, paired males were redder than males without a mate (Hill 1990). Perhaps the most convincing evidence comes from manipulations of color in the wild. Hill (1991) brightened 40 males with hair dyes, lightened another 40, and sham-treated 20, all before breeding commenced. Of the males that were re-sighted, 96% of the brightened males paired with a female, compared to 60% of the controls, and 27% of the lightened males. Preferences of females for redder males seem to be general over time and among populations, and are found whether redness is rated subjectively by humans or with a reflectance spectrophotometer (Hill et al. 1999).

RELIABILITY OF CAROTENOID COLORATION

What could carotenoid pigmentation signal that females would benefit from knowing? One possible answer is current condition, in the sense of the present nutritional state of the male. Data clearly demonstrate that diet affects carotenoid pigmentation, showing that a link exists between nutrition and carotenoid levels. Addition of carotenoids to the diet increases red or orange coloration in guppies (Kodric-Brown 1989, Grether 2000) and in house finches (Brush and Power 1976, Hill 1992, 1994). If we assume that the amount of carotenoid obtained in the diet in nature is proportional to how much food has been eaten, and thus to the amount of energy and nutrients obtained, then these results imply that a male's color should indicate something about his nutritional state. Hill (2000) demonstrated another link between carotenoids and nutrition by manipulating access to food independently of access to carotenoids in captive house finches. Males that were periodically deprived of food while molting were less red after molt than were control males that were not food-deprived, even though all were given excess carotenoids in their water. This result again argues that coloration signals nutritional state. Note, however, that plumage color in finches changes only at molt, which typically precedes mate choice by several months, so the information on nutritional state conveyed by color may be severely out of date by the time females make their assessment.

Although the link between diet and carotenoid levels implies that carotenoids might signal nutritional state to females, direct evidence requires showing a correlation between carotenoid level and male nutritional state at the time females exercise mate choice. The best evidence here comes from fish, whose color can fluctuate over much shorter time periods than that in birds. Ichthyologists typically measure nutritional condition as mass divided by length raised to the power b, where b is the slope of the regression of log mass on log length. Milinski and Bakker (1990) found a significant correlation between this "condition factor" and the intensity of red coloration in captive male sticklebacks (figure 3.3a). Frischknecht (1993), however, found no such correlation in a second sample of captives, nor was a correlation found by Bakker and Mundwiler (1994) among a sample of free-living males. Candolin (1999) used lipid concentration to measure condition in free-living sticklebacks, and found a U-shaped relationship between area of red and condition—the males with the most red were either in very good or very bad condition (figure 3.3b).

The current health of an individual, especially its parasite load, is another aspect of male quality that might be signaled by carotenoids. Experimental exposure to parasites has been shown to lower carotenoid pigmentation in several species. In sticklebacks, for example, exposure to an infectious ciliate was followed by a significant decrease in red coloration in males (Milinski and Bakker 1990). In male guppies, spots of carotenoid pigment became paler and less saturated after exposure to a monogenean parasite, whereas the spots of control males did not change (Houde and Torio 1992). In house finches, males experimentally exposed to coccidia molted into plumage that was significantly less red than the plumage of control males, as measured by a spectrophotometer (Brawner et al. 2000). Inadvertent infection with mycoplasma also had a negative effect on red coloration in this study. Experimental infection with

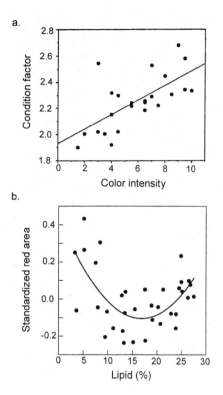

FIGURE 3.3. Two examples of relationships between condition and color in three-spined sticklebacks. **a.** A significant positive correlation between the intensity of the red breeding coloration of males and their condition factor (weight corrected for body length), from Milinski and Bakker (1990). **b.** A U-shaped relationship between the area of red (standardized for body length) and lipid content as a percentage of dry weight, from Candolin (1999).

coccidia lowered absorption of carotenoids in chickens (Ruff et al. 1974, Tyczkowski et al. 1991).

The fact that exposure to parasites lowers levels of carotenoid pigmentation implies that carotenoid colors should be inversely related to levels of parasitism, but complications are rife. One problem is that carotenoids may themselves be important in immune defenses against parasites (Lozano 1994). As Shykoff and Widmer (1996) point out, if carotenoids must be allocated either to pigmentation or to immune defense, and all individuals have roughly equal access to carotenoids, then those individuals that allocate more generously to pigmentation will have lower levels of defense and higher levels of infestation, creating a positive correlation between color and parasite levels. If access to carotenoids, rather than allocation, varies among individuals, then those with especially good access can afford high levels of both defense and display, and a negative correlation between color and parasites is expected. Further complications arise if individuals vary in resistance to parasites independently of carotenoid allocation, if the food items that are a source of carotenoids are also vectors for parasites (Folstad et al. 1994), and if the levels of different parasites vary independently of each other (Weatherhead et al. 1993).

Empirical data are mixed on the relationship between carotenoid pigmentation and parasite levels; positive relationships are observed in some studies and negative relationships in others (Shykoff and Widmer 1996). In male sticklebacks, for example, Folstad et al. (1994) found significant positive correlations between intensity of red coloration and levels of infestation for two cestode parasites and a significant negative correlation for a third; this third species may be the most pathogenic. In great tits, the hue of carotenoid-pigmented feathers (measured in this case with a photospectrometer) was lower in males infected with blood parasites than in uninfected males among yearlings, but the reverse was true for older males—infected males had higher hue values (Hõrak et al. 2001). In house finches, males with few feather mites molted into redder plumage than did males with many mites, among both yearling and older males (Thompson et al. 1997). Many other studies have found no relationship at all between carotenoid coloration and parasites (Shykoff and Widmer 1996).

Even when carotenoid coloration does reflect male health, females still ought to choose males on color only if male health somehow affects female fitness. One obvious way this could occur is if healthy, well-fed males provide better parental care than poorly fed, parasite-ridden ones. Indeed, Hill (1991) found that the rate at which male house finches feed their young is positively correlated with the brightness of their plumage. Similarly, Candolin (2000) showed for sticklebacks that the survival of eggs correlated positively with the extent of red on the father caring for them. Another possibility is that carotenoid pigmentation advertises good genes for foraging ability or parasite resistance, and thus for viability. Hill (1991) found that more-colorful male house finches have higher overwinter survival. Similarly, Nolan et al. (1998) found that those male house finches that survived an outbreak of mycoplasmal conjunctivitis were redder than those that succumbed. These results suggest that redness does signal viability in this species, but whether these viability effects are heritable is unknown. A third possibility is that a female benefits directly from avoiding parasitized males because of the danger that she herself will become infected.

Rodd et al. (2002) have suggested that the preference shown by female guppies for carotenoid colors in males may be explained by sensory bias, a hypothesis which if accepted would make questions of reliability moot. The evidence supporting sensory bias is that guppies appear to be attracted to orange fruit in the wild, and when tested with disks of differing colors approach and peck at orange and red disks in preference to disks of other colors. In between-population comparisons, the strength of the female preference for orange males is strongly correlated with the rate at which both females and males peck at orange disks. Rodd et al. (2002) therefore argue that female preference for orange in males may be a nonselected consequence of visual biases evolved in the context of foraging. If so, the apparent benefits females obtain from responding to carotenoids in males, for example from mating with males in better condition or with fewer parasites, would be no more than coincidental to the evolution of female response. We have difficulty accepting such a coincidence. To date there is little evidence that the female preference in guppies evolved before the male display trait, which is usually taken as the key test of a sensory-bias hypothesis. In the absence of such evidence we think it more likely that female response to orange coloration in male guppies has evolved because of the information that the signal provides on mate quality, as usually assumed.

COSTS OF CAROTENOID PIGMENTATION

When carotenoids honestly signal male quality, we expect to find some cost enforcing that honesty. One constraint on the expression of carotenoids is their rarity in the diets of some animals. If carotenoids are sufficiently rare in an animal's diet, then the amount of carotenoids accumulated and displayed can be an index of how much food has been ingested, and thus of the animal's nutritional condition. Rarity of carotenoids is a constraint rather than a cost, but it will be associated with an opportunity cost if animals must spend more time foraging in order to gather sufficient carotenoids for display. This argument assumes that the amount of carotenoids used in display is limited by the amount in the diet; a limitation of this sort has often been shown in captivity, but only rarely in nature. As one example, nestling great tits obtain yellow carotenoid pigments, at least in part, from Lepidopteran larvae fed to them by their parents (Slagsvold and Lifjeld 1985). Such larvae are more abundant in deciduous forest than in coniferous forest, and Slagsvold and Lifjeld (1985) found that the saturation of yellow in the plumage of young great tits increased with the relative amount of deciduous foliage around the nest. In addition, nestlings with above-average yellow were found to have been fed higher proportions of Lepidopteran larvae than were those with below-average yellow. Similarly, Grether et al. (1999) found that carotenoid availability varies between streams in the natural habitat of guppies. Among streams, the mean amount of carotenoids found in the foreguts of guppies was positively correlated with the concentration of carotenoids in their color spots. Thus the amount of carotenoid obtained in the diet was directly related to the amount used in display. Finally, Hill et al. (2002) measured carotenoid levels in the gut contents of molting house finches taken from two populations that differed in the extent of red plumage. The population with more red plumage was found to have over three times as much carotenoid in the gut as the population with less red plumage. Within populations, carotenoid levels in the gut were positively correlated with plumage brightness. Thus all three studies indicate that carotenoid availability constrains pigment ornamentation in nature, though it may be that carotenoids are more widely available for other species in other habitats (Hudon 1994, Thompson et al. 1997). Further investigation of carotenoid limitation in nature clearly is needed.

Another possible cost of using carotenoids in display stems from the direct health-enhancing effects of these compounds (Lozano 1994). Carotenoids have an array of positive effects on health (Olson and Owens 1998), through their action as antioxidants (von Schantz et al. 1999) and as stimulants of the immune system (Lozano 1994). Carotenoids deposited as pigments in dead tissues such as feathers are not recoverable for health-related uses, though pigments may be recoverable if deposited in living tissues, such as wattles and combs in birds or skin spots in fish. Regardless of whether allocation is permanent or temporary, if carotenoids are needed for combating parasites and disease, then their allocation to display is costly, and is an honest signal that the individual is in good health, or was in good health when the allocation was made.

Hill (1999) has argued against a health cost of carotenoid pigmentation, on the grounds that species with carotenoid displays have much higher levels of circulating carotenoids than species lacking such displays. Species with carotenoid displays thus may have more than enough carotenoids still available for health-related functions. Several experiments, however, have shown that supplementing dietary carotenoid enhances immune response even in species that use carotenoids for display. Grether et al. (2004) manipulated the level of carotenoids in the diet of captive guppies, and found a stronger rejection of foreign scales in males given a high-carotenoid diet than in males given a low-carotenoid diet. Two other experiments were done with zebra finches, a species that uses carotenoids in the bill and legs for sexual display. Males given additional carotenoids had both stronger cell-mediated responses (Blount et al. 2003, McGraw and Ardia 2003) and stronger humoral responses (McGraw and Ardia 2003) to foreign antigens compared to males whose dietary carotenoids were not supplemented. Contrasting results were obtained in American goldfinches (*Carduelis tristis*), a sexually dichromatic species in which males exhibit extensive, carotenoid-based yellow coloration. The level of carotenoids provided in the diet during the pre-alternate molt affected plumage color, but had no effect on either humoral or cell-mediated immune response (Navara and Hill 2003). Evidence for the health cost of carotenoid display thus remains mixed, at least for birds.

A third possible cost of carotenoid-based displays is a metabolic one, involving the cost of processing and mobilizing the pigments needed to produce the display (Hill 2002). Carotenoids that are ingested, say by a house finch, must at a minimum be transported in the blood by carrier proteins, which themselves cost something to synthesize and maintain (Hill 2000). Further, it often is the case that the chemical form of the pigment actually used for display differs from the form that is ingested, requiring metabolic processing before the pigment is ready to be deposited in the feathers, again at some cost. That such costs as these can limit coloration is shown by the experiments in which nutrition affects pigmentation independently of access to dietary carotenoids (Hill 2000).

A fourth possible cost is an increased risk of predation, due to bright, carotenoid coloration attracting the attention of predators. In both sticklebacks and guppies, variation in color pattern between localities is related to the distribution of predators, with drabber color patterns predominating where visually hunting predators are present (Moodie 1972, Endler 1983). In laboratory experiments, predatory trout were more likely to attack sticklebacks artificially colored with red lipstick than to attack controls to which clear lipstick had been applied (Moodie 1972), and predatory blue acara cichlids (*Aequidens pulcher*) were more likely to attack the more brightly colored of two individuals when presented with guppies that varied naturally in their expression of red carotenoid pigments (Godin and McDonough 2003). Endler (1980, 1983) showed in a greenhouse experiment that a particularly voracious predatory fish, *Crenicichla alta*, selected for a reduction in the size of red spots in guppies. In a field experiment, transferring a guppy population away from *Crenicichla* led to an increase in the size of red spots in subsequent generations (Endler 1980). These selection experiments provide unusually direct evidence that the signal has fitness costs, at least in fish. The case is different, however, in birds. When young great tits were painted red, they experienced increased predation as expected (Götmark and Olsson 1997), but when adult European blackbirds (*Turdus merula*) were given red wing patches, the experimental birds were actually attacked significantly less often than the controls (Götmark 1994, 1996). Although these experimental manipulations of plumage provide less direct evidence than the selection studies done with fish, it seems clear that a predation cost of red coloration is not universal.

The various costs of carotenoids are not mutually exclusive and may, in fact, be additive. If carotenoids are difficult to obtain, useful for fighting disease, and expensive to transport and modify, this may make it triply difficult for an animal in poor condition to produce a colorful display. Although evidence demonstrating costs is mixed in some cases, there is sufficiently strong evidence overall to conclude that various costs and constraints, either alone or in concert, are sufficient to maintain the honesty of carotenoid colors as signals of condition.

Deceit in Carotenoid Coloration

Although there is evidence that carotenoids can be reliable indicators of male quality, it is still possible that at times carotenoids are used deceitfully. The best evidence for deceit via carotenoids comes from Candolin's (1999, 2000) work with sticklebacks. Candolin (2000) measured coloration in male stickle-backs, first when held in isolation and then when held in groups of four. The mean area of red exhibited per male decreased significantly when males were grouped, while variation among individuals in area of red increased. Measures of hue and intensity of coloration did not change. The decrease in area of red among interacting males implies that there may be some receiver-dependent cost of the signal, via attracting aggression from other males (see chapter 4). Area of red correlated with male parental ability (measured as the survival of eggs under male care), and the correlations were significantly stronger when area of red was measured after interaction than when measured before interaction. That area of red decreased after interaction and was then a better predictor of male quality strongly implies that before interaction some males were exaggerating their quality, and were in that sense being dishonest.

In a separate study, Candolin (1999) found a U-shaped relation between lipid content and a standardized measurement of the area of red coloration in male sticklebacks (figure 3.3b). Lipid content was used as a more direct measure of male condition than the usual ratio of mass to size. The fit of the quadratic was highly significant ($r^2 = 0.36$, P < 0.001), whereas a linear regression showed a very poor fit ($r^2 = 0.03$, P > 0.3). Note that the males with the largest areas of red were among those in poorest condition. Candolin (1999) went on to show that males that were experimentally deprived of food for an extended period increased their red areas relative to control males that were not food-deprived. The difference in color between experimentals and controls increased when both sets of males were exposed to predators, that is, when they could see two perch (*Perca fluviatilis*) in another tank. The effect occurred because the food-deprived males showed less of a reduction in area of red during exposure to the predators than did the control males.

In interpreting these results, Candolin (1999) assumed that the value of the signal to females is mainly as a predictor of paternal care. This seems a reasonable assumption, given that the display correlates with quality of paternal care (Candolin 2000) and that the quality of a direct benefit such as parental care is particularly important to female fitness (Maynard Smith 1991b). Candolin also assumed that lipid reserves directly affect the quality of care that a male can provide. From these premises, it follows that males with low lipids and large areas of red are signaling dishonestly. If the chances of later reproduction are low enough for a male in very poor condition, it may benefit him to pay the cost of exaggerating his signal now, since he will not live to reproduce again anyway, as predicted in the optimal allocation model of Kokko (1998). This rationale also explains why males in poor condition continue to signal strongly under increased risk of predation—they accept the increased risk because they do not have much to lose. As in the models of Kokko (1997, 1998), the overall signaling system can still be stable if the dishonest males are rare enough that the system is "honest on average," in the sense that females are more likely to obtain a high-quality mate by attending to the signal than by choosing randomly.

Songs in Oscine Birds

> "All those who have attended to the subject, believe that there is the severest rivalry between the males of many species to attract by singing the females." (Darwin 1859, *On the Origin of Species*, 1st Edition, pp. 88–89)

Songs can be defined in general as long, complex vocalizations produced mainly in the breeding season (Catchpole and Slater 1995). Vocalizations that meet this simple definition can be found in a number of animal groups, including most frogs and toads, certain insects, whales, and primates, as well as several orders of birds. The best-known singers are the oscine birds, a subset of the order Passeriformes (passerine birds or "perching birds"), defined taxonomically by the complex musculature of their vocal organ, the syrinx. Although a great deal is known about the function of song in many of those other groups (Gerhardt and Huber 2002, Greenfield 2002), we confine ourselves in this section to the songs of oscine birds.

In most temperate species of oscines, songs are produced exclusively, or nearly exclusively, by males. Singing by females seems to be considerably more common in tropical species than in temperate ones (Stutchbury and Morton 2001), and given the diversity of tropical avifaunas, many more oscines are tropical than are temperate. Nevertheless, we will treat song as a male phenomenon, simply because the available data on song as a signaling system comes almost exclusively from temperate species where male song predominates.

Male song is thought usually to have dual functions, one in male-male aggression, and the other in attracting and courting females. Both functions apply in oscines (Catchpole and Slater 1995) as well as other taxa (Searcy and Andersson 1986). Here we will concentrate on the mate attraction and courtship function, but the possibility of male-male aggressive effects must be kept in mind when interpreting certain types of results, such as correlations between song features and mating success.

RECEIVER RESPONSE TO BIRD SONG

Song is a complex behavior that varies dramatically between species, and as a consequence a great many song features exist to which female oscines might respond (Searcy and Nowicki 2000). Most of the features for which there is evidence of female response can, however, be assigned to one of four categories: song output, song complexity, local song structure, and vocal performance.

SONG OUTPUT

Song output refers to the amount of song produced by a male, and is usually measured as song rate or song duration. Correlations between male song rate and pairing success in free-living birds suggest that females prefer males that sing at the highest rates. In the village indigobird (*Vidua chalybeata*), for example, males sing from exposed perches and are visited by females for mating. Among a variety of male and location characteristics, Payne and Payne (1977) found song rate to be the best predictor of the number of mates obtained per male. In the monogamous willow warbler (*Phylloscopus trochilus*), Radesäter et al. (1987) found that male pairing date was negatively correlated with song rate; thus males with the highest song rates were chosen first. Song rate was also correlated with male age, but the correlation between pairing date and song rate remained significant when age was controlled. In the laboratory, Houtman (1992) found that female zebra finches preferred to associate with males having high song rates, and that association preferences predicted whether a male would be chosen for extra-pair copulation. Females also preferred males with redder beaks, and beak redness was correlated with song rate, but when song rate and beak color were both entered in a multiple regression, only song rate predicted attractiveness. In a later study, Collins et al. (1994) experimentally manipulated beak color, and found that female zebra finches preferred males with fast song rates and dull beaks over ones with slow song rates and bright beaks.

Although these studies of female preference controlled for some male characteristics other than song rates, we can never be sure that correlational studies such as these have identified and controlled all the important characters other than the one of interest. What are needed are experimental studies in which the trait of interest is manipulated. Experimental studies of song rates have been rare; an exception is the study of pied flycatchers by Alatalo et al. (1990). Previous work had shown that males with naturally high song rates paired earlier in this species (Gottlander 1987). Alatalo et al. (1990) manipulated song rates by providing males with mealworms on their territories. Experimental males sang more than twice as much as did unsupplemented control males. In matched pairs of experimental and control males, the experimental male attracted a female before the control in 11 of 13 comparisons (P < 0.05). Again, a female preference for high song output is indicated, although it remains possible that females responded to some other behavior that was changed by the feeding regime but was not measured by the researchers.

SONG COMPLEXITY

Song complexity is most often measured as song repertoire size, which in turn may be measured either as the number of different syllable types that a male possesses, as in *Acrocephalus* warblers, or as the number of song types, as in song sparrows (*Melospiza melodia*). A variety of evidence suggests a fairly general preference of female oscines for larger repertoires and thus for greater complexity. Some of the evidence comes from laboratory tests in which females were first treated with estradiol, to prime them to perform courtship displays, and were then exposed to playback containing varying numbers of syllable types or song types. For example, we found, using this paradigm, that female song sparrows gave more solicitation display in response to four song types than to one, more for eight types than for four, and more for 16 types than for eight (Searcy and Marler 1981, Searcy 1984). In a similar way, female great tits responded preferentially to larger over smaller repertoires of song types (Baker et al. 1986), and female sedge warblers (*Acrocephalus schoeno- baenus*) and great reed warblers (*A. arundinaceus*) to larger over smaller reper- toires of syllable types (Catchpole et al. 1984, 1986).

Correlational evidence from the field supports a female preference for larger repertoires in some cases, but intercorrelations between repertoire size, age, and territory quality can make causality hard to pinpoint. For example, in a pioneering study of northern mockingbirds (*Mimus polyglottos*), Howard (1974) found that pairing date was negatively correlated with both repertoire size and a measure of territory quality. When territory quality was controlled, repertoire size was no longer correlated with pairing date. One interpretation of these results is that repertoire size affects success in male-male competition for territories, and success in competition for territories in turn affects pairing. In the highly polygynous red-winged blackbird, Yasukawa et al. (1980) found positive correlations between repertoire size and both pairing success and age. When age was controlled, the correlation between repertoire size and pairing success disappeared, suggesting that the relationship between pairing success and repertoire size might have been only an indirect consequence of an effect of age on pairing. Recently, Reid et al. (2004) have shown that, for first-year male song sparrows that have acquired a territory without a resident female, the probability of attracting a mate increases with repertoire size.

Support for a direct effect of repertoire size on female choice comes from studies of two species of *Acrocephalus* warblers. In the socially monogamous sedge warbler, Catchpole (1980) found that syllable repertoire size was nega- tively correlated with date of pairing; that is, males with larger repertoires obtained mates earlier. In a later study on the same species, Buchanan and Catchpole (1997) confirmed the relationship between repertoire size and pair- ing date across three different breeding seasons, and found that the correlation was maintained when measures of song output and territory size were con- trolled. In a long-term study of the polygynous great reed warbler, Hasselquist (1998) found that repertoire size was not correlated with the number of social mates acquired when male age was controlled. Repertoire size was, however, a good predictor of success in obtaining fertilizations outside the pair bond, better than either male age or territory quality (Hasselquist et al. 1996). In ten of ten comparisons, the male obtaining an extra-pair fertilization had a larger repertoire than the male he cuckolded (Hasselquist et al. 1996).

Lampe and Sætre (1995) provide a convincing experimental confirmation of female preferences for larger repertoires in pied flycatchers under seminatural conditions. In an aviary, a female was presented with two males in separate compartments. Next to each male's compartment was a nestbox, with a loud-speaker on top. The researchers played a large song repertoire from one speaker and a small repertoire from the other. Females expressed their choice of mates by building a nest in one of the nestboxes. Only seven of the 12 females tested actually built a nest, but all seven of those building a nest did so in the nestbox associated with the larger repertoire.

LOCAL SONG STRUCTURE

Song in oscine birds often, perhaps always, exhibits geographic variation within species. Geographic differences are easiest to discern in those species in which individuals sing only one or a very few song types. Here, all the males in a given area may sing the same song type or types, and the boundaries between adjacent song traditions can be both well-defined and abrupt. Classic examples of such "dialect" systems are provided by the white-crowned sparrow (*Zonotrichia leucophrys*) in North America and the corn bunting (*Miliaria calandra*) in Europe (Marler and Tamura 1962, McGregor 1980). At the other extreme are species in which each male sings multiple song types and not all song types are shared between adjacent males. In these systems, songs are likely to change gradually with distance rather than abruptly at discrete boundaries, and the precise nature of between-population differences is difficult to perceive against the background of within-population variation. Nevertheless, evidence suggests that geographic variation does occur in these more complex song systems, as for example in chaffinches (*Fringilla coelebs*) (Slater et al. 1984) and song sparrows (Borror 1965).

Female oscines show a seemingly ubiquitous preference for songs originating in their own, local area over foreign songs of the same species. Good evidence of such a preference exists for the white-crowned sparrow, one of the classic dialect species. Baker (1983) captured 20 female white-crowned sparrows from the "Clear" dialect in central California. The females were about one month old at capture and were tutored in captivity with their natal dialect for another two months. At six months the females were put on long days, treated with estradiol, and then tested for response to songs of their natal dialect and songs of the adjacent, "Buzzy" dialect. The females gave three times as many solicitation displays in response to their home dialect as to the foreign dialect. Additional tests with white-crowned sparrows, using other combinations of home and foreign dialects, have obtained the same result: females respond more strongly to home dialect than to foreign dialects (Baker 1983, Baker et al. 1981, 1987).

We have studied female preferences for local song in a species with a more complex song system, the song sparrow. Each song sparrow male sings six to 16 quite distinct song types, and each of these song types contains many different note types (Podos et al. 1992). Males in our study population in the eastern United States rarely share songs with neighbors (Hughes et al. 1998), and the diversity of song patterns within a locale is thus much greater than the already considerable diversity within individual males. Given the variability of song within locales, we have difficulty discerning any consistent differences between localities; nevertheless, the birds themselves discriminate readily between local and foreign songs. We have tested female song sparrows for discrimination of songs taken from a transect running approximately 540 kilometers, from Hartstown, in northwestern Pennsylvania, to Millbrook, in southeastern New York (Searcy et al. 1997, 2002). Taking Hartstown as the starting point, we recorded songs at various points along the transect as well as at the end points. We then tested captive, estradiol-treated females from Hartstown for response to local versus foreign songs. Hartstown females failed to discriminate between local songs and songs recorded at the 1/32 point, but gave significantly more solicitation displays for local songs than for songs from the 1/16, 1/8, and 1/4 points, and than for songs from the opposite (Millbrook) endpoint (figure 3.4). As Baker (1983) found for white-crowned sparrows, the preferences were quite strong: an approximately fourfold difference in the number of displays given to local songs versus songs from the opposite endpoint (540 km distant), and a twofold difference in displays given to local songs versus songs from the 1/16 point, just 34 km distant.

Although results from both song sparrows and white-crowned sparrows provide unequivocal evidence that females respond preferentially to local song relative to foreign song, neither study demonstrates that the preferences for local song actually affect mate choice. Evidence that mating decisions are influenced by song preferences is provided by a series of studies of the brown-headed cowbird (*Molothrus ater*) by Meredith West, Andrew King, and colleagues. The brown-headed cowbird is a brood parasite in which males do not hold territories, which eliminates territory quality as a possible confounding influence on female mate choice. King et al. (1980) found that captive female cowbirds of two subspecies, *Molothrus ater ater* and *M. a. obscurus*, were more responsive to male songs of their own subspecies than to songs of the other subspecies. The *ater* females and songs were taken from Maryland, and the *obscurus* females and songs from Texas. Females of the *ater* subspecies gave solicitation displays in response to 49% of *ater* songs, compared to only 33% of *obscurus* songs. Females of the *obscurus* subspecies showed the opposite preference, as expected, responding to 18% of the *ater* songs and 36% of the *obscurus* songs. When males and females of both subspecies were held together in aviaries, pairing was significantly more likely to occur between birds of the same subspecies than between birds of different subspecies (East-

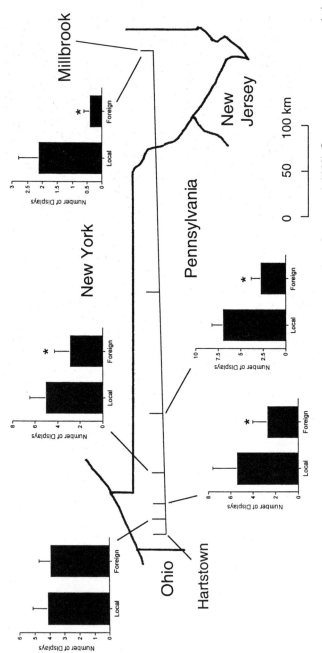

Figure 3.4. Female preferences for local song over foreign song in song sparrows (from Searcy et al. 2002). Song sparrows were recorded at a series of sites along a transect running from Hartstown, Pennsylvania (on the west) to Millbrook, New York (on the east). The histograms show mean response of Hartstown females to local songs versus foreign songs recorded at some of the intermediate points. Females did not discriminate between local songs and foreign songs from the first foreign site, 18 km from Hartstown, but significant discrimination was shown against foreign songs recorded at the second foreign site, 34 km from Hartstown. From this site on, discrimination tended to grow stronger with increasing distance of the foreign site from the local site.

zer et al. 1985). The best evidence that the song preferences cause the pairing preferences comes from experiments with "bilingual males"—that is, ones that sing the songs of both subspecies. West et al. (1983) produced bilingual males by housing juvenile males of one subspecies, with adult males of the other subspecies, or with adult males that were themselves bilingual. Subsequently, when bilingual males were housed with *ater* females, each male's copulatory success was positively correlated with the proportion of *ater* songs that he sang; conversely, when bilingual males were housed with *obscurus* females, each male's copulatory success was positively correlated with the proportion of *obscurus* songs he sang (West et al. 1983).

Freeberg and colleagues have studied a second pair of cowbird populations, a *Molothrus a. ater* population from Indiana and a *M. a. artemesiae* population from South Dakota. Freeberg (1996) raised South Dakota juveniles with either South Dakota or Indiana adults; these juveniles can then be termed South Dakota-culture and Indiana-culture individuals, respectively. When housed together in mixed groups, South Dakota-culture females paired preferentially with South Dakota-culture males, and Indiana-culture females with Indiana-culture males. Freeberg et al. (1999) raised a second generation of South Dakota juveniles with either the South Dakota-culture or the Indiana-culture birds; this second generation also mated preferentially with others of the same culture. These experiments indicate that the two populations must differ in behavior, that the differences are passed from generation to generation by learning, and that these behavioral differences are important to mating. Exactly how behavior differed was not clear from these studies, but Freeberg et al. (2001) went on to show that Indiana males differed from South Dakota males in the fine structure of their songs, and that the songs of Indiana-culture and South Dakota-culture males differed in parallel ways. Finally, captive females of the South Dakota population gave significantly more solicitation display in response to playback of South Dakota-culture songs than to Indiana-culture songs. Thus females definitely respond preferentially to songs of their own locality, and it seems highly likely that the song preferences are at least in part responsible for the observed patterns of mating.

VOCAL PERFORMANCE

We can assume that physical and physiological constraints exist on what kinds of sounds songbirds can make and how rapidly and loudly they can make them (Nowicki et al. 1992, Podos 1997, Suthers and Goller 1997, Podos and Nowicki 2004). "Vocal performance" refers to the ability of individual males to push such limits. Other aspects of male quality may be correlated with the ability of males to approach limits on vocal performance, making vocal performance of interest to females. Since we are only beginning to learn what limits exist on song production, we are only beginning to be able to measure performance in meaningful ways. Nevertheless, a few examples of female response to song performance have begun to emerge.

Vallet and Kreutzer (1995) showed that female canaries are especially responsive to a particular phrase type, termed an "A phrase." Females respond preferentially to A phrases no matter where in the song these phrases are positioned (Vallet and Kreutzer 1995). A phrases are characterized by the rapid repetition of a two-note syllable, with both notes frequency modulated. Testing with a variety of A phrases showed that female canaries especially preferred those A phrases having the most rapid repetition rates (Vallet et al. 1998). Vallet et al. (1998) suggest that these complex two-note phrases are difficult to produce, and that females prefer rapid versions of these phrases because these demonstrate superior vocal performance.

Dusky warblers (*Phylloscopus fuscatus*) produce trills of short syllables, each containing one or more frequency-modulated notes. Forstmeier et al. (2002) measured vocal performance for these songs as the proportion of the trill during which the sound amplitude exceeded 20% of the song's peak amplitude. They suggested that this measure is affected by how rapidly a male can refill his air sacs between syllables, which has been suggested to limit the speed of sustained trills in other species (Suthers and Goller 1997). Forstmeier et al. (2002) calculated the residual performance for a male by taking the difference between his performance on a given syllable type and the population mean performance and then summing these differences over all the syllable types produced by that male. Residual performance was consistently correlated with extra-pair paternity: males that obtained extra-pair paternity had higher residual performance than the males they cuckolded in seven of seven comparisons. Males chosen as extra-pair males had higher residual performance than available males that were not chosen in five of five comparisons. Residual performance was not, however, correlated with success in attracting social mates. The evidence thus suggests that females attend to vocal performance in choosing extra-pair mating partners but not in choosing long-term social mates.

The best understood example of a limit to vocal performance concerns the relationship between the frequency bandwidth and repetition rate of trilled syllables. In order to produce a high-frequency sound, a male bird must open his bill widely, which shortens the vocal tract and raises its resonance frequency; conversely, to produce a low-frequency sound the male must close down the bill, lengthening the vocal tract and lowering its resonance frequency (Nowicki 1987, Westneat et al. 1993, Podos et al. 1995, 2004, Hoese et al. 2000). Producing a syllable with a wide frequency bandwidth therefore requires opening and closing the bill over a wide angle. Assuming there is a limit to how fast a bird can open and close its bill, these factors necessarily produce a tradeoff between bandwidth and repetition rate: the wider the bandwidth of a syllable, the lower is the maximum rate at which it can be produced (Podos 1997).

Evidence for the existence of this trill rate/bandwidth constraint comes from song-learning experiments with swamp sparrows (*Melospiza georgiana*). The songs of swamp sparrows consist of frequency-modulated syllables repeated in rapid trills. Podos (1996) tutored young male swamp sparrows with songs whose trill rates were artificially increased. The males were unable to duplicate the faster trill rates, resorting to expedients such as simplifying the notes or introducing gaps in the trills in order to circumvent the problem. These results support the existence of a limit to the speed at which syllables of a given structure can be sung. Additional evidence comes from a comparative analysis of the songs of the New World sparrows (Emberizidae), the family that includes swamp sparrows. Pooling songs from many species of sparrows, Podos (1997) found a triangular distribution for bandwidth plotted against trill rate (figure 3.5a). Syllables having a narrow bandwidth were produced at either fast or slow rates, but syllables with wide bandwidths were produced only at slow rates. An upper-bound regression, which focuses on the maximum bandwidths observed for categories of trill rate, gives an estimate of the location of the performance limit. Deviation of songs from the upper-bound regression can be used to measure the vocal performance of that song—the closer to the line, the better the performance (figure 3.5b).

Ballentine et al. (2004) found that a large sample of songs from a swamp sparrow population also showed a triangular distribution of bandwidth versus trill rate (figure 3.5c). The upper-bound regression for this single population was statistically indistinguishable from that found for the sparrow family as a whole. Female swamp sparrows tested with pairs of songs, one near and one far from the upper bound, consistently performed more courtship displays for the songs close to the performance limit (figure 3.5d). Females again seem to prefer songs of higher vocal performance.

RELIABILITY AND COSTS OF SONG FEATURES

As we have discussed previously, a number of different aspects of male quality might be communicated, reliably or otherwise, by signals such as song features. In addition, female preferences for song features can evolve by a couple of mechanisms without those features signaling male quality in any way. Finally, what type of cost is most relevant to reliability may depend on which aspect of quality is being signaled and which song feature is doing the signaling. In an attempt to organize this complex set of interacting factors, we will consider in turn the issues of reliability and cost for each of the categories of song features we have outlined.

RELIABILITY OF SONG OUTPUT

The amount of song a male produces per unit time might depend directly on his energy balance, thus making song output a signal of male condition.

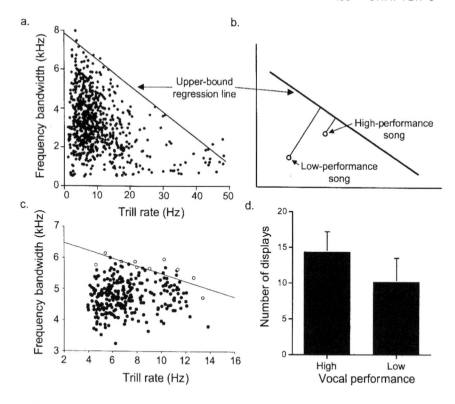

FIGURE 3.5. Vocal performance and female preferences in New World sparrows (Emberizidae). **a**. The triangular distribution of frequency bandwidths and trill rates from a comparative analysis of sparrow species (from Podos 1997). Songs were binned by trill rates in 5-Hz increments, and the maximum bandwidth for each bin was regressed against trill rate. The resulting "upper-bound regression" line gives an estimate of the limit to vocal production. **b**. A schematic illustrating how vocal performance is measured as the orthogonal deviation of a song from the upper-bound regression. **c**. The distribution of frequency bandwidths and trill rates from an analysis of 280 songs from a single population of swamp sparrows (from Ballentine et al. 2004). The upper-bound regression line found for this sample is statistically indistinguishable from that found for the Emberizidae as a whole. **d**. The response of female swamp sparrows to swamp sparrow songs of high and low vocal performance (from Ballentine et al. 2004). The preference for high-performance songs is significant ($P < 0.05$).

Energy balance might in turn depend on the availability of food on the male's territory, so that song output could also signal territory quality. The best evidence that song rates signal food availability comes from experiments in which males were provided with extra food on their territories and the effect on song rates was monitored. As an example, Davies and Lundberg (1984) placed food on the territories of randomly chosen male dunnocks (*Prunella modularis*), starting in January and extending well into the breeding season. Early in the breeding season, during the period in which females were most likely to settle, 15 provisioned males sang at a mean rate of 126.4 songs per hour, more than double the 56.7 songs per hour produced by 19 control males. Provisioned males also commenced singing earlier in the season. Similar, if not always so dramatic, increases in song rates have been observed in response to provisioning in red-winged blackbirds (Searcy 1979), Carolina wrens (*Thryothorus ludovicianus*) (Morton 1982), and pied flycatchers (Gottlander 1987). High song output thus should be a reliable signal to a female that food is plentiful on a potential mate's territory.

Females who choose males in possession of abundant food may benefit from their own easy access to food, or they may benefit through obtaining a mate who is himself well-fed and therefore able to do a superior job in caring for young. In willow tits (*Parus montanus*), the song output of males was significantly positively correlated with their total effort in feeding the young (Welling et al. 1997). In addition, males that attacked a model of a nest predator had higher song output than males that did not attack. Similar relationships were found by Grieg-Smith (1982) in stonechats (*Saxicola torquata*), with song in the pre-breeding period predicting a male's later contribution to both nest guarding and feeding of nestlings.

If male song rate signals either food availability or the quality of a male's future parental care, then a female choosing a male with a high song rate ought to benefit directly by producing additional young. In neither the willow tit nor the stonechat, however, was song output positively correlated with the production of offspring (Grieg-Smith 1982, Welling et al. 1997). The expected correlation was found in a study of blackcaps (*Sylvia atricapilla*) by Hoi-Leitner et al. (1993); here the number of young raised per female was positively correlated with male song rate. But this correlation did not come about because high song output predicted better paternal care; on the contrary, male participation in incubation and feeding of the young was strongly negatively related to male song rates. Nor was there any direct evidence that females choosing males with high song rates benefited from obtaining more food on the males' territories. Instead, a subsequent study (Hoi-Leitner et al. 1995) showed that male song rates correlated with vegetation density on the territory, and that vegetation density was related to the probability that nests escape predation. In a partial correlation analysis, female preferences were found to be better related to male song rate than to vegetation density, so it appears that females were choosing social mates on the basis of song rate rather than directly on territory quality.

These studies with blackcaps illustrate the difficulties of disentangling the motivations for female preferences. Even when a direct benefit is demonstrated for the preference, various possibilities remain for how that benefit comes about. The hypothesis offered by Hoi-Leitner et al. (1995) is that safety from nest predation (rather than food) is the important determinant of territory quality, that males with high song rates are (for whatever reason) successful in competition for good territories, and that females choose on song rates rather than on territory quality because vegetation has not developed seasonally when females make their choices. Thus, male song rate does correlate with direct benefits a female might receive, but the link between the preferred trait and the benefit is quite circuitous.

It is also possible that females obtain indirect, genetic benefits, rather than direct, material benefits, from preferring males with high song output. To demonstrate indirect benefits, one must show that song output in a male predicts viability in his offspring. The closest anyone has come to establishing this relationship is Houtman's (1992) demonstration in zebra finches that a father's song rate correlates with his offsprings' weight at independence. The weight of a young bird at independence has been shown in many cases to correlate with its future survival (Gebhardt-Henrich and Richner 1998). Thus Houtman's results are consistent with the view that a male's song output predicts some aspect of his genetic quality related to growth, that the father's genetic quality affects offspring genetic quality, and that offspring genetic quality in turn affects offspring growth and survival. On the other hand, the observed correlation between a father's song output and the weight of his offspring might equally well come about because the father's song output predicts the quality of paternal care, as in willow tits and stonechats.

Another aspect of genetic quality that might be signaled by song output is parasite resistance. Møller (1991a) showed that song output in male barn swallows was negatively correlated with mite abundance at their nests. To test causality in this relationship, Møller (1991a) manipulated mite abundance by adding mites to some nests and killing mites at others with insecticide. Song output subsequently dropped precipitously in males of the mites-added treatment relative to controls, but was not much affected in the mites-lowered treatment. The mites infecting barn swallows take blood from both adults and young, and may be transmitted between individual birds (Møller 1994a). Because resistance to mites is heritable (Møller 1990a), females might receive a genetic benefit from preferring mite-free males. Alternatively, females might benefit from avoiding mite-infested males because of the possibility of transmission of the mites to themselves or their young (Møller 1994a). Garamszegi et al. (2004) demonstrated that the song output of collared flycatchers (*Ficedula albicollis*) decreased when males experienced an immunological challenge in the form of an injection of sheep red blood cells. This result is again consistent with the view that song output reflects something about the current health and energy budget of an individual, although it remains to be seen if this effect reflects heritable differences among males.

COSTS OF SONG OUTPUT

Song output in males signals a number of kinds of information that might be of interest to females, including the abundance of food on a male's territory, the quality of the male's future parental care, and the male's condition in terms of energy balance and parasite load. Which of these is of most interest to females, and why, remain open questions. Regardless, it seems reasonable that exaggerating any of these qualities might well be advantageous to males, so we are again confronted with the problem of what costs enforce reliability.

If singing had a sufficient energy cost, only males that were well-fed and unburdened by parasites could afford to sing at high rates for long periods. Thus, given the types of information signaled by song output, it seems logical to suspect that reliability is maintained by an energy cost of song production. Song definitely has some energy cost, but whether this cost is high enough to enforce honesty is another matter. A standard way to estimate the energy cost of a behavior is to measure the amount of oxygen consumed while performing that behavior, which varies as a function of metabolic rate. Eberhardt (1994), who was the first to attempt such measurements on a singing bird, found that oxygen consumption by singing Carolina wrens increased by a factor of about 3.9 relative to resting metabolic rate (RMR, measured during sleep), suggesting a substantial energy cost for song. Subsequent studies have failed to find similarly high metabolic costs, however. Oberweger and Goller (2001) measured oxygen consumption during singing in individuals of three oscine species. In all three species, oxygen consumption increased during singing relative to RMR, but only by a factor of 2.0 in zebra finches, 2.6 in wasserslager canaries, and 2.2 in European starlings. Ward et al. (2003) measured the cost of song in fife fancy canaries and roller canaries, two breeds differing from each other and from wasserslagers in song characteristics, and found factorial increases of 2.5 and 2.1 over RMR, respectively. Finally, Ward et al. (2004) measured oxygen consumption during song in wild pied flycatchers, finding in this case a factorial increase of 2.7 over RMR.

These costs, though not staggering, seem appreciable, but using energy costs during sleep as the baseline inflates the apparent cost of song, because energy consumption is higher in awake birds regardless of whether they are singing. A more realistic estimate of the cost of song comes from comparing oxygen consumption during song to oxygen consumption of the bird standing (without moving) immediately prior to song (Oberweger and Goller 2001). Considered this way, the metabolic costs associated with singing appear quite unimpressive, increasing by a factor of 1.28 in zebra finches, 1.12 in pied flycatchers, 1.07 in fife fancy canaries, 1.05 in roller canaries, 1.04 in wasserslager canaries, and 1.06 in starlings. Eberhardt's (1994) data from Carolina wrens still remain on the high end, with a 3.6-fold increase above metabolic rate while standing (see table 2 in Ward et al. 2004), although this discrepancy may be due to the fact that the wrens were more active in their metabolic chambers and that standing metabolic rates were not measured immediately prior to singing, thus inflating the estimate of oxygen consumption during song (Oberweger and Goller 2001). All in all, the bulk of the available evidence suggests that the energy cost of singing, although greater than nothing, is too low for it to put an obvious limit on song output.

Even if the direct energy expense of singing is low, energy may still put an indirect limit on song output through opportunity costs. In many species of birds, singing and foraging are incompatible activities, so a bird choosing to sing must forgo foraging, if only temporarily. In savannah sparrows (*Passerculus sandwichensis*), for example, Reid (1987) found that males either sang while sitting still or foraged while moving about. Males that sang during a half-hour observation period foraged significantly less than males that did not sing. By forgoing foraging, a bird loses the opportunity to increase its overall energy budget. Whether such an opportunity cost is important to a bird must depend on how badly it needs energy. This brings us back to studies showing that males provided with food on their territories sing more (e.g., Searcy 1979, Morton 1982, Davies and Lundberg 1984, Gottlander 1987), a result that can only mean that male birds during the breeding season are limited in how much they can sing by some combination of time and energy. Thus, the cost of decreasing foraging opportunities seems the most likely mechanism for enforcing honesty in song output, but more evidence is needed to strengthen this conclusion, particularly evidence on how singing affects overall energy budgets.

RELIABILITY AND SONG COMPLEXITY

Song complexity has very different properties as a signal than song output, because it is so much less labile. Song output can vary from minute to minute within an individual, whereas some aspects of song complexity may be set at an early age and never change thereafter. In song sparrows, for example, the song type repertoire is complete when a male first begins to sing crystallized song at the age of one year, and song types are neither added nor subtracted for the remainder of the individual's lifetime (Nordby et al. 2002). In other species, such as the red-winged blackbird, repertoire sizes remain constant within years, but song types can be added between years (Yasukawa et al. 1980). Other aspects of song complexity can change more rapidly; for example, both song sparrows and red-winged blackbirds vary the frequency with which they switch between song types, depending on the context (Kramer and Lemon 1983, Searcy and Yasukawa 1990). Nevertheless, the two most frequently used measures of complexity, song repertoire size and syllable repertoire size, typically are fixed at least for the duration of a breeding season, if not for an entire lifetime.

How song complexity can signal something about a male that is of interest to a female is not obvious at first glance, but some studies have demonstrated significant relationships between complexity and certain measures of male quality. In sedge warblers, for example, Buchanan and Catchpole (2000) found the syllable repertoire size of males to be significantly correlated with the rate at which they fed their young. The weight of the young, corrected for age, also

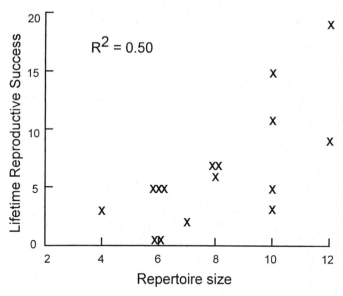

FIGURE 3.6. The correlation between song repertoire size and lifetime reproductive success for a song sparrow population on Mandarte Island, British Columbia (from Hiebert et al. 1989). Variation in song repertoire size predicts an amazing 50% of the variation in lifetime reproductive success.

increased with male repertoire size. In European starlings and pied flycatchers, however, the amount of male parental care was not related to repertoire size (Mountjoy and Lemon 1997, Rinden et al. 2000). In the sedge warbler, repertoire size was negatively related to the presence of blood parasites (Buchanan et al. 1999), such that parasitized males exhibited significantly smaller repertoires than parasite-free males. In pied flycatchers, repertoire size was positively correlated with male body weight and with an index of condition based on weight corrected for body size (Lampe and Espmark 1994).

The most impressive evidence that song complexity can reliably signal something about male quality comes from studies relating repertoire size and lifetime reproductive success (LRS). Hiebert et al. (1989) measured song repertoire sizes and LRS in a cohort of 16 male song sparrows on Mandarte Island in British Columbia. The correlation between repertoire size and LRS was strong—amazingly strong—with repertoire size accounting for 50% of the variation in LRS (figure 3.6). Repertoire size also was correlated with the number of months that a male held territory, suggesting that the relationship with LRS might come about in part because song complexity was somehow related to longevity, or because it had an effect on male-male competition for territory. Repertoire size, however, was still correlated with LRS when territory tenure was controlled, and was also strongly correlated with annual reproductive success, a component of fitness unaffected by tenure. Song repertoire size also has been found to be correlated with lifetime production of offspring in great tits (McGregor et al. 1981) and great reed warblers (Hasselquist 1998).

Two kinds of explanations can account for these observed correlations between fitness and repertoire size. One is that repertoire size has a direct effect on fitness, through effects on female choice, male-male competition, or both. The other is that repertoire size is only indirectly related to fitness through correlations with other male traits that affect survival or reproduction, in other words with male quality. Direct effects of repertoire size on fitness are possible, for experimental evidence indicates that repertoire size does affect female choice in song sparrows, great tits, and great reed warblers (Searcy 1984, Baker et al. 1986, Catchpole et al. 1986). The evidence also indicates that repertoire size directly affects male-male competition for territory in great tits (Krebs et al. 1978). Nevertheless, it seems doubtful that direct effects of repertoires on fitness can be the complete explanation, or even the larger part of the explanation, because it seems so unlikely that any one character, and especially any one display character, could on its own determine half the variation in fitness in any species, as would be required in song sparrows. Instead, it seems far more likely that repertoire size is an indicator of other male traits that are directly important to fitness, and that the correlation of repertoire size with fitness comes about largely because of its relationship with these other aspects of male quality. Direct effects of repertoire size on fitness would still be possible, but they would be subsidiary to this indirect relationship.

COSTS OF SONG COMPLEXITY

Many of the types of costs that apply to other signals involved in mate choice seem unlikely to apply to song complexity. A large repertoire of song types should require no more energy to produce than does a small repertoire. In their measurements of oxygen consumption during song, Ward et al. (2003) found no difference between roller canaries and fife fancy canaries, although the latter breed produces much more complex songs as measured by the number of different phrases included. If anything, Lambrechts and Dhondt (1988) have argued, just the opposite could be the case: switching among song types or phrase types might reduce muscular fatigue. Nor does it seem likely that a large repertoire would decrease immunological function or increase predation risk. A more reasonable cost for song complexity is suggested by the "developmental-stress hypothesis" (Nowicki et al. 1998, Buchanan et al. 2003b, Nowicki and Searcy 2004). Under this hypothesis, song complexity is determined in part by song-control nuclei in the brain, and the development of these nuclei imposes costs at a period when the young bird is both particularly vulnerable to nutritional stress and rapidly developing its phenotype. Only those individuals that happen to escape stress or that have genotypes particularly resistant to stress can devote the resources needed to develop complex song while simulta-

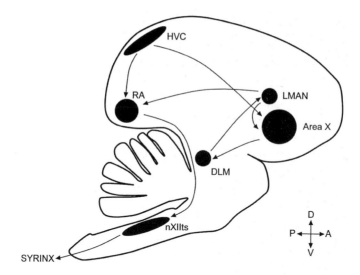

FIGURE 3.7. The neural pathways known to affect the production and development of song in songbirds. Two main pathways are recognized. The motor pathway, which controls song production, runs from the HVC to the RA (both in the forebrain) to the nXIIts (a motor nucleus in the brainstem) to the syrinx, which produces the sounds. The learning pathway, which is important in song development, runs from the HVC to Area X, to DLM, to LMAN, to RA. See Brenowitz and Kroodsma (1996).

neously completing other aspects of development. The reliability of song complexity as an indicator of male quality thus is ensured by the developmental costs associated with the ability to produce complex repertoires.

Learning, storage, and production of song are mediated in oscine birds by two series of interconnected brain nuclei (figure 3.7). The motor pathway starts with HVC, leads then to RA, the tracheosyringeal portion of the hypoglossal nucleus (nXIIts), and the syrinx. Lesions of HVC or RA lead to severe deterioration of adult song (Nottebohm et al. 1976). A second set of nuclei, the "anterior forebrain loop," connects HVC with Area X, DLM, and LMAN, and then links back to the motor pathway at RA. Lesions in Area X or LMAN are especially detrimental to song acquisition in juveniles, rather than to song production in adults (Bottjer et al. 1984, Sohrabji et al. 1990).

The size of HVC, which is a nexus in both pathways, has been linked to song complexity by comparisons at a number of levels. The original evidence comes from a study by Nottebohm et al. (1981), who found a significant, positive correlation between HVC volume and repertoire size in a sample of 25 male canaries. Subsequently, positive relationships between HVC size and repertoire size have also been found within populations of marsh wrens (*Cistothorus palustris*), zebra finches, and sedge warblers (Canady et al. 1984, Airey and DeVoogd 2000, Airey et al. 2000). No such relationship was found, how-

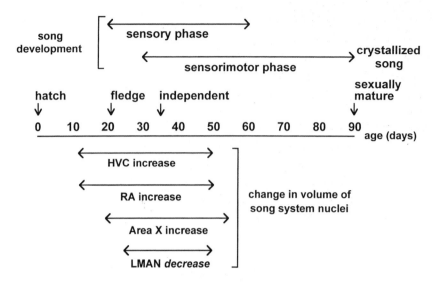

Figure 3.8. The timing of development of song and the song system in zebra finches, from Nowicki et al. (1998). The sensory phase of song development is the period over which model songs are memorized, and the sensorimotor phase is the period over which songs are practiced, leading to the final (crystallized) song at 90 days. The sensory phase in particular overlaps widely with the period of growth in volume of important song-control nuclei, such as HVC, RA, Area X, and LMAN.

ever, in red-winged blackbirds, spotted towhees (*Pipilo maculatus*), or European starlings (Kirn et al. 1989, Brenowitz et al. 1991, Bernard et al. 1996). In a between-populations comparison, Canady et al. (1984) found that western marsh wrens had both larger repertoire sizes and larger HVC volumes than eastern birds of the same species. Finally, at the interspecific level, DeVoogd et al. (1993) used the independent-contrasts method to compare HVC volumes (controlled for telencephalon size) and repertoire sizes in 41 species of oscines while controlling for phylogenetic relationships. Contrasts in HVC volumes were significantly and positively correlated with contrasts in repertoire size. A similar study by Székely et al. (1996) found a significant positive relationship between HVC and repertoire size among a more focused and better controlled sample of nine warblers in the family Sylviidae.

The development of the song-control system has been described more thoroughly in the zebra finch than in any other bird (Bottjer et al. 1985, Nordeen and Nordeen 1988). In this species, the HVC and RA increase in volume during the period 10 to 50 days after hatching, and Area X increases in volume 20 to 50 days after hatching (figure 3.8). Interconnections between the song-control nuclei grow somewhat earlier, 10 to 35 days post-hatching (Nowicki et al. 1998). Nordeen et al. (1989) examined song-system development in swamp sparrows, a more typical songbird, and found the timing of song-system development in this species to be remarkably similar to that observed in zebra finches, with the majority of growth of HVC, RA, and Area X completed by 61 days post-hatch.

These periods of brain development overlap substantially with the memorization phase of song learning, when the young bird listens to the song production of older males and stores acceptable song models for later use. In zebra finches, this phase extends from 25 to 65 days post-hatching (Immelmann 1969, Slater et al. 1988). In song sparrows and swamp sparrows, most song memorization appears to occur 10 to 50 days post-hatching (Marler and Peters 1987, 1988). Although many factors may influence the precise timing of song acquisition in different species, the general pattern is that the first two or so months are particularly important in determining the outcome of a young bird's vocal development. This interval corresponds to the latter part of the nestling period, the first part of the post-fledging period when young birds are still dependent on their parents for food, and the first few weeks after they become independent of their parents. These are all periods during which young birds are highly likely to experience developmental stress, especially stress resulting from poor nutrition. Energy demands are high during the nestling stage, owing to rapid growth (Ricklefs 1974), and at least in some cases are even higher after fledging (Martin 1987). Starvation is common among both nestlings and fledglings in altricial birds (O'Connor 1984, Ringsby et al. 1998). Starvation may be even more likely early in the period of complete independence (Weathers and Sullivan 1989) because of the poor foraging skills of the young (Sullivan 1988). Stresses other than nutritional ones are also likely early in life; in particular, parasites are often more common on nestlings and juveniles than on adult birds.

Growth in mass, skeletal features, and feathers is rapid during the nestling period, and in many species continues after fledging (O'Connor 1984). Post-hatch growth rates often vary considerably among individuals in a population, owing at least in part to external factors affecting food availability, such as clutch size, weather, or quality of parental care (Ricklefs and Peters 1979, 1981, Ricklefs 1983). Effects of diet on growth during these early periods can have permanent effects on the bird's phenotype. Boag (1987), for example, showed that the protein content of the diet affected growth rates in zebra finch young, and that the differences set up in this way carried over into adulthood for a variety of skeletal features. We have demonstrated recently that a brief exposure to poor nutrition occurring over an approximately 2-week period after hatching results in pronounced skeletal size differences in song sparrows that persist throughout adulthood (Searcy et al. 2004)

The effects of nutrition on brain development have been little studied in birds, but are well known in mammals, where nutritional deficits early in life have been shown to produce permanent effects on the size of brain areas and associated learning abilities (Smart 1986, Levitsky and Strupp 1995). Studies of the mammalian hippocampus are particularly relevant, for this structure, like the song-control areas of birds, is directly tied to specific behavioral abilities, having to do in the case of the hippocampus with spatial learning and memory. In laboratory rats, poor early nutrition negatively affects numbers of cells, cell size, and dendritic branching in the hippocampus (Castro and Rudy 1987, Levitsky and Strupp 1995).

To summarize, the developmental-stress hypothesis suggests that the complexity of song is determined in part by the development of the song-control nuclei, that these nuclei grow during a period when the young bird is also developing other important aspects of its phenotype, and that the development of the song-control system and development of the overall phenotype are both vulnerable to the nutritional deficits and other stresses that are likely during this period. Stresses that harm phenotypic development will also harm song development, and vice versa. Song complexity therefore becomes a signal of the overall quality of a male's phenotype, and the reliability of the signal is maintained by the developmental cost of building a brain suitable for the task of learning a complex song.

An obvious prediction of the developmental-stress hypothesis is that the song complexity of adults should reflect their developmental history as young. This prediction has been confirmed in a field study of great reed warblers, one of the species in which song complexity has been shown to be correlated with lifetime reproductive success (Hasselquist 1998). Nowicki et al. (2000) found a positive, nearly significant relationship between syllable-repertoire size of adult great reed warblers and their body mass as nestlings. Syllable-repertoire size of adults was also positively correlated with nestling feather length, and this relationship was significant. These results show that song development does correlate with general phenotypic development, and the quality of learned song thus contains information about how well an adult fared during early development.

A recent experimental study also provides evidence of an effect of early developmental stress on song complexity. Buchanan et al. (2003b) stressed young European starlings during the period roughly 40 to 120 days post-hatching by removing their food for 4 hours per day. Control young had uninterrupted access to food. As adults, the stressed group sang shorter song bouts (Buchanan et al. 2003b) and had significantly smaller repertoire sizes (Spencer et al. 2004) than the controls. Bout duration and repertoire size are strongly, positively correlated in starlings (Eens et al. 1991), and both song features are positively correlated with female preferences (Eens et al. 1991, Mountjoy and Lemon 1996).

The developmental-stress hypothesis also predicts that early nutrition should affect the development of the song-control system and song learning. We have tested these predictions in a laboratory experiment with swamp sparrows

(Nowicki et al. 2002a). Nestling swamp sparrows were taken from the field and hand-reared under two nutritional regimes, a control regime in which the subjects were given unlimited food, and an experimental regime in which the subjects were restricted to 70% of the food intake of the controls. The restricted-food group grew more slowly in mass and tarsus length than did the control group. As adults, the nutritionally restricted group had significantly smaller HVC volumes, RA volumes, and telencephalon volumes than did the controls. The two groups did not differ in the ratio of HVC volume to telencephalon volume, but the ratio of RA volume to telencephalon volume was significantly lower in the nutritionally restricted birds. Thus nutritional deprivation early in life had permanent effects on the song-control nuclei. We did not find any effect of the nutrition treatments on repertoire size of adults, perhaps because swamp sparrows show little variation in the sizes of their repertoires. The nutrition treatments did, however, affect the accuracy with which the young swamp sparrows copied tutor songs. In this experiment, then, nutrition affected the quality of learning rather than the quantity. We suggest that effects of nutrition on quality of learning may also explain female preferences for local song, the subject of the next section.

RELIABILITY AND COST OF LOCAL SONG STRUCTURE

As we reviewed earlier, female songbirds commonly prefer songs produced by local males over songs produced by conspecific males residing at more distant sites. The distant sites need not be very distant, as in song sparrows, whose females prefer local songs over songs from males living just 34 kilometers away. The acoustic differences between geographic variants in song are often rather subtle, and therefore not of a nature that could possibly make one variant more costly to produce than another. Moreover, female preferences seem to be reciprocal, in the sense that females from site A prefer site A songs over site B songs, whereas females from site B show the opposite preference. Hence, even if songs from one site are somehow more costly to produce than songs from the other, this would not provide an explanation for the general pattern of female preferences. Production of local song thus seems unlikely to be a signal of male condition, and in this respect resembles song complexity rather than song output.

If local song structure does not communicate male condition, what information might this signal contain that could explain why females pay attention to it? Three hypotheses have been proposed to account for this preference: (1) that geographic variation in song reveals local genetic adaptation on the part of males; (2) that female preferences for local song are a nonselected by-product of selection against mating with heterospecific males; and (3) that adherence to local song structure is a test of the quality of song learning and hence of the quality of the male's phenotype, as suggested by the developmental-stress hypothesis.

Nottebohm (1969, 1972) originally proposed that mating within local populations might enhance adaptation to local ecological conditions in birds, and that dialect systems function to facilitate local mating. These suggestions were made in the context of explaining the evolution of song learning and song dialects, rather than in an attempt to account for female preferences, which at the time were unknown. In fact, the local genetic-adaptation hypothesis predicted the occurrence of female preferences for local song, which were later demonstrated. The plausibility of the local adaptation idea depends in part on when song learning occurs relative to dispersal. If males learn songs after leaving their natal area, then song obviously works less well as a marker of place of origin. The timing issue has been much debated, especially for the white-crowned sparrow (*Zonotrichia leucophrys*), which has been used extensively as a model species to test the local-adaptation hypothesis (Kroodsma et al. 1985). Song memorization from tape recordings closes off gradually between 50 and 100 days of age in white-crowned sparrows (Marler 1970), and tutoring from live males may (Baptista and Petrinovich 1986) or may not (Cunningham and Baker 1983, Nelson 1998a) extend the sensitive period. Regardless of the effect of social tutoring, it seems likely that some memorization occurs after dispersal, given that dispersal can start as early as 35 days. Memorization after dispersal also seems likely in other species of songbirds (Baker and Cunningham 1985). Nevertheless, song may still act as a fairly good marker of natal origin, as long as dispersing males do not move very far.

Another controversial point in the white-crowned sparrow saga has been whether song dialects serve as markers of genetic differences. Baker et al. (1982) found differences in allozyme frequencies among a series of dialect populations of white-crowned sparrows found in coastal California (*Z. l. nuttalli*). This discovery provoked a debate over whether genetic differences between dialects were greater than expected by distance alone (Zink and Barrowclough 1984, Baker et al. 1984). This debate seems to us somewhat beside the point, in the sense that as long as both song and gene frequencies change with distance, song differences will contain some information about genetic differences, even if there are no boundaries where both song and gene frequencies change abruptly. Work with neutral genetic markers (microsatellites) suggests that white-crowned sparrow populations in the Sierra Nevada (*Z. l. oriantha*) show significant variation in allele frequencies among dialects, though the proportion of the total variation ascribable to among-dialect differences (0.79%) is tiny compared to the proportion due to individual differences within populations (98.70%) (MacDougall-Shackleton and MacDougall-Shackleton 2001). More recently, however, Soha et al. (2004) failed to find either significant genetic divergence among song dialects in Puget Sound populations of white-crowned sparrows (*Z. l. pugetensis*) or a correlation between genetic distance and geographic distance in this subspecies. These authors also reanalyzed the data from Baker et al. (1982), using improved statistical techniques.

Although they found a correlation between genetic distance and geographic distance among these coastal California populations, they found no relationship between genetic distance and dialect after controlling for geographic distance. The best evidence currently available, then, suggests little or no relationship between dialect and the genetic characteristics of white-crowned sparrow populations, or even between genetic distance and geographic distance in this species. Nonetheless, results from tests of female response indicate that song does change increasingly with distance in white-crowned sparrows (Baker 1983), as is true of the song sparrows that we study (Searcy et al. 2002).

Although we do not consider that either the timing of song learning or the pattern of genetic change pose fatal problems for the local genetic-adaptation idea, we do see a more important problem in the lack of evidence for local adaptation in birds. Local adaptation requires that individuals be better adapted genetically to conditions at their home site than to conditions at other sites, while individuals at the other sites show reciprocal adaptation. Many of the songbirds that show geographic variation in song, such as white-crowned sparrows and song sparrows, occupy similar habitat over wide areas, so that there would seem to be little selection for adaptive genetic differences, at least not at the scale of tens of kilometers, at which changes of song and song preferences are seen. Those genetic differences that are observed may well be neutral, rather than adaptive. To our knowledge, only one study has suggested a possible advantage of locally adapted genes in white-crowned sparrows. Mac-Dougall-Shackleton et al. (2002) showed that birds singing the local dialect in a population had lower parasite loads than birds singing a foreign dialect. This result is consistent with local males having better genetic adaptations to deal with the local parasites, though it could also be explained by differences in prior exposure to the local parasites or by a difference between dispersing and nondispersing individuals in the overall quality of their immune systems (MacDougall-Shackleton et al. 2002).

Another critical difficulty with the genetic-adaptation hypothesis is that dispersal may often be limited enough that females may never encounter males whose songs are discriminably different from those of local males. In song sparrows, for example, females do not discriminate against songs recorded at a distance of 18 kilometers, but do discriminate against songs from a distance of 34 kilometers (Searcy et al. 2002). The root mean square dispersal distance for song sparrows has been estimated to be in the range of 0.35 to 6.1 kilometers, depending on the estimation method used (Barrowclough 1980, Zink and Dittman 1993). The root mean square dispersal distance represents the standard deviation of the distance moved from an individual's natal site. Given that the dispersal distance is so much less than the distance at which songs can be discriminated, it must be very rare indeed for a female to encounter a male whose songs she can discriminate from local songs. The ability to reject foreign males therefore seems unlikely to be an important benefit of female preferences for local song structure.

A second possible explanation of female preferences for local song is that these preferences are a nonselected by-product of selection against mating with heterospecific males. Selection against hybrid matings apparently has led to reinforcement of prezygotic isolating mechanisms in many taxa (Coyne and Orr 1997, Rundle and Schluter 1998), including birds (Sætre et al. 1997). Species song recognition is an important isolating mechanism in birds, and selection might well favor strengthening such recognition. Nelson (1998b) has suggested that songbirds achieve species song recognition by memorizing songs heard from conspecifics early in life, and by using these memories in adulthood as an internal standard against which to judge songs heard during mate choice. Because the songs heard and memorized early in life are local songs, this mechanism of species recognition creates a bias against foreign songs, even if this bias is not itself of any advantage. The problem we see with this hypothesis is that it does not explain the extreme specificity of song preferences seen in many species of songbirds. In song sparrows, for example, the differences between the songs of nearby song sparrow populations are minute (Searcy et al. 2003) compared to the differences between song sparrow songs and the songs of even closely related species (Nowicki et al. 2001), and yet female song sparrows show preferences based on the minute, between-population differences. This level of discrimination ability is overkill as far as species recognition is concerned, and to us demands some other explanation.

The explanation we favor is the developmental-stress hypothesis. Under this hypothesis, females use the accuracy with which males have copied local song as a measure of the quality of song learning. Quality of learning matters to females because it reflects the developmental experience of males early in life, when many aspects of their phenotype are developing. The hypothesis predicts that stresses experienced early in life should affect both general phenotypic development and the accuracy of song learning, which brings us back to the experiment in which we nutritionally stressed a group of young swamp sparrows and measured the effects on song development (Nowicki et al. 2002a). Again, the subjects for this experiment were taken from the field as nestlings and hand-reared. Half the subjects, the controls, were given unlimited food, while the other half, the experimentals, were restricted to 70% of the food ingested by the controls. All the subjects were tutored with tape recordings of the same set of swamp sparrow songs. Accuracy of learning was measured by spectrographic cross-correlation (Clark et al. 1987) between the spectrograms of the learned notes and the model notes. The songs produced by the nutritionally stressed group were significantly poorer copies of the models to which they had been exposed than were the songs produced by the control group. Thus, as predicted, nutritional stress early in life had a negative effect on the accuracy of song learning.

In a separate experiment, we tested whether differences in the accuracy of learning, of the sort measured in the above experiment, actually affect female preferences (Nowicki et al. 2002b). In this experiment, we used female song sparrows, rather than swamp sparrows, as test subjects. The test songs were ones recorded from male song sparrows hand-reared in the lab and tutored with songs recorded from our study population in Hartstown, Pennsylvania. We rated the songs produced by the young males on two measures of learning: the proportion of notes copied from the model songs, and the accuracy of copying as indicated by spectrogram cross-correlations between model and copied notes. Well-learned copies, which rated well on both measures of learning, were just as acceptable to wild-caught females as were the model songs, originally recorded from free-living males (figure 3.9a). We also tested the females for discrimination between a set of well-learned copies and a set of poorly learned copies; these two sets of songs differed in both the proportion of notes copied and the average copy accuracy. Females showed strong discrimination in favor of the well-learned songs and against the poorly learned ones (figure 3.9b). Finally, we tested the females for discrimination between a set of well-learned songs and a set of copies that rated well on the proportion of copied notes but poorly on the average copy accuracy. The females discriminated against these latter songs, of intermediate learning quality, but the discrimination was not as strong as in the previous comparison (figure 3.9c).

In these female preference tests, subjects were taken from the same local population as were the songs used to tutor males in the learning experiment. Thus the local standard for what a song ought to sound like should have been the same for females as for the hand-reared males. Females discriminated against songs that departed from this standard, in the sense either that the singer had failed to learn many notes or that the notes he learned were copied inaccurately, illustrating how females can use adherence to local song structure as a test of a male's song-learning performance. By the developmental-stress hypothesis, the cost that maintains reliability of local song structure as an indicator of male quality is, again, the cost of developing the brain mechanisms necessary for young male birds to learn accurately the features of those songs.

RELIABILITY AND COSTS OF PERFORMANCE FEATURES

Because the study of vocal performance is in its infancy, not much is yet known about the reliability of performance features. The general expectation would be that as vocal performance is limited by factors such as muscular coordination and respiratory capacity, vocal performance might be correlated with success at other tasks that also are demanding in coordination and endurance, such as flight or foraging. The one positive piece of evidence that bears on this prediction is that singing performance was found to correlate positively with overwinter survival in dusky warblers (Forstmeier et al. 2002). Curiously, good vocal performance was associated with losing rather than winning territo-

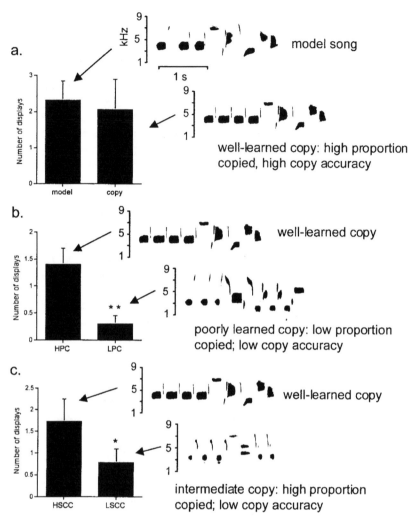

FIGURE 3.9. The response of female song sparrows to the songs of hand-reared males tutored with songs recorded in the female's local population, from Nowicki et al. (2002b). Female response is measured as the number of copulation-solicitation displays performed in response to playback. Ten examples of each category of songs were used in the experiments. **a.** Females showed no discrimination between a set of the model songs (recorded from free-living males) and a set of well-learned copies, which had both a high proportion of copied notes and a high average note-copy accuracy. **b.** Females showed strong discrimination in favor of well-learned copies and against poorly learned copies, which had a low proportion of copied notes and low note-copy accuracy. **c.** Females showed somewhat weaker discrimination in favor of well-learned copies and against a set of copies of intermediate quality, well-learned in having a high proportion of copied notes, but poorly learned in having low note-copy accuracy.

rial fights; Forstmeier et al. (2002) suggest that this association might be due to allocation decisions, whereby males unable to sing well enough to attract females invest more heavily in aggression, at some cost to survival.

If performance features do turn out to be reliable predictors of male quality, that reliability might be explained by extending the developmental-stress idea. Just as investment in brain development is required for young birds to learn more complex songs and to copy accurately the idiosyncratic features of songs that identify local dialects, so does the development of peripheral neural and motor mechanisms represent a cost that must be paid if a bird is to have a vocal apparatus equal to the task of producing high-performance songs. Birds that differ in the amount of stress they incur early in life, or in their ability to cope with this stress, will differ in the amount they can invest in the development of peripheral structures associated with song production. Future work is needed to test these ideas.

Tail Length in Birds

Darwin (1871) noted that in some species of birds the tails of males were elongated relative to those of females, sometimes greatly so, and he gave some anecdotal evidence that tail length is important to female choice of mates in such species. Over the next 100 years, tail length in birds remained a favorite example of a sexually selected trait, likely to be adduced whenever a hypothetical example was needed of how sexual selection might work. Therefore, it was highly appropriate that Malte Andersson chose to do his pioneering work on manipulating ornamental traits on just this character. The resulting study (Andersson 1982) was immensely influential in convincing other biologists of the reality of female preferences for purely ornamental traits, that is, traits that have no natural selective advantage.

Since this first manipulative study, experimental evidence has appeared for a number of species showing that females prefer longer tails, making this one of the better documented cases for receiver response to a courtship character. Tail length in birds has also been intensively, if not extensively, studied from the point of view of reliability and costs, thanks especially to work done on barn swallows and blue peafowl (*Pavo cristatus*). Accordingly, we address tail length here as a third category of mating signals.

RECEIVER RESPONSE TO TAIL LENGTH

Andersson's (1982) classic study was carried out on the long-tailed widowbird (*Euplectes progne*), a species in which the tails of adult males are about 50 cm long, compared to only 7 cm in females. The mating system of this African species is territorial polygyny, meaning that individual males hold territories

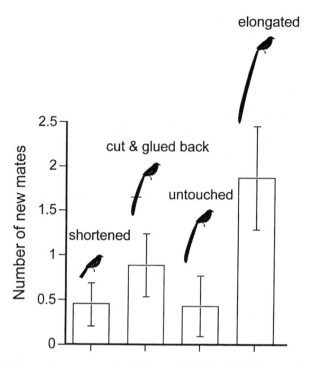

FIGURE 3.10. The mating success of male long-tailed widowbirds after experimental alteration of their tail lengths, from Andersson (1982). Shortened males had their tails cut to ~14 cm. The removed feathers were glued onto the tails of elongated males, adding ~25 cm. One set of controls had their feathers cut and then glued back on; the tails of the second set were not manipulated. The number of new mates was estimated from the number of nests built by females on male territories after the manipulations. Males with elongated tails were preferred over all other groups.

on which multiple females nest. Andersson's method, much followed by subsequent researchers, was to shorten the tails of one set of males by cutting their feathers, and then to use the cut feathers to lengthen the tails of another set of males. These manipulations reduced tail length in the shortened males to about 14 cm, and increased tail lengths in the elongated males to about 75 cm. Andersson used two control groups, one in which the tail was cut and then glued back, and the other unmanipulated. The number of new nests started on a territory after treatment was used as the measure of female preference. The result was a significant increase in female preference from shortened to control to elongated males (figure 3.10). The enhanced success of the males with elongated tails was particularly apparent.

The tail of male long-tailed widowbirds is, by any standard, greatly exaggerated, but even more impressive, and more widely familiar, is the tail of the blue peafowl. It is actually the upper tail coverts, rather than the tail feathers proper, that are elongated in peacocks; the resulting "train" is spread and displayed both when courting females and in aggressive encounters with other males. Peacocks hold display territories on leks, which females visit for mating. Working with a feral population in England, Petrie and colleagues showed a positive correlation between train size (measured by the number of eye-spots) and the number of matings obtained by males (Petrie et al. 1991, Petrie and Halliday 1994). Experimental manipulation of eye-spot number had a somewhat ambiguous effect on male mating success (Petrie and Halliday 1994). The clearest evidence for female preferences came from observing the sequence of males visited by individual females leading up to mating; 10 of 11 females chose to mate with the one male out of those they visited with the greatest number of eye-spots, and the remaining female chose a male with just one eye-spot less than the maximum she had encountered (Petrie et al. 1991).

The tail of the male barn swallow is a less spectacular ornament than those of widowbirds and peafowl, but this rather low-key ornament nevertheless has an effect on female preferences. The outer tail feathers of barn swallows are elongated in both sexes relative to most other swallows, and these "streamers" are further elongated in males (to about 11 cm) relative to females (about 9 cm) (Møller 1994a). Barn swallows are socially monogamous, so female preferences are expressed more subtly than in polygynous and promiscuous species. Møller (1990b) examined natural variation in tail length in a Danish population, and found that males with longer tails paired more quickly than males with shorter tails. Møller (1988b) manipulated tail lengths, using the same types of treatments as Andersson's (1982) widowbird study had, and found that males with elongated tails paired earlier than controls, and males with shortened tails paired later. Smith and Montgomerie (1991b) subsequently confirmed the effects of tail manipulation on pairing date in an Ontario population, but Safran and McGraw (2004) failed to find this relationship in a population in New York State. Two studies have found that females mated to males with elongated tails are less likely to be fertilized by other males (Møller and Tegelström 1997, Saino et al. 1997b), whereas a third study found the opposite (Smith et al. 1991). One of these studies showed that males with elongated tails were more likely to fertilize females other than their social mates (Saino et al. 1997b). Thus, although there appears to be some variation among populations, male tail length has an influence on the mating preferences of female barn swallows on a number of levels—at the level of their choice of a social mate, at the level of deciding whether or not to copulate with that social mate, and at the level of choosing which extra-pair males to copulate with.

Overall, the above studies have been highly successful in showing the effects of tail length on mating success, either using correlations of natural tail length with mating success (see also Palokangas et al. 1992, Regosin and Pruett-Jones 2001, Pryke et al. 2001) or through experimental manipulation of tails (see also Andersson 1992). These studies provide strong evidence that sexual selection is acting, but they provide only weak evidence for a signaling function of the tail—that is, of a direct, proximate response of receivers to the signal. A better study in this latter regard is that of Mateos and Carranza (1995) on tail length in ring-necked pheasants (*Phasianus colchicus*). Here, two stimulus males were placed in individual cages, separated by a partition. One male's tail was artificially lengthened while the other male served as an unmanipulated control. Midway through the experiment the treatments were reversed, so that the elongated male became the control and vice versa. Females were tested singly for preference, the criteria being that they spend > 75% of the time in proximity of one male while also courting him at least once. Overall, 19 of 25 (76%) of those females showing a preference chose the male that at the time had the longer tail.

RELIABILITY OF TAIL LENGTH

The reliability of tail length, as a signal of benefits that the male will provide to the female, has been investigated more thoroughly in barn swallows than in any other species. Møller and colleagues have shown that male barn swallows with long tails are phenotypically superior in various respects; for example, they tend to be older (Møller 1994a), to have higher hematocrits (Saino et al. 1997a), and to have fewer fault bars in their tail and wing feathers (Møller 1989b). Female barn swallows might benefit directly from mating with superior males, perhaps by obtaining better territories or parental help for their offspring. Territories in barn swallows are tiny, however, and apparently of not much importance to mate choice (Møller 1994a). Male barn swallows do provide considerable parental care (Møller 1994a), but the quality of paternal care, in terms of the amount of food brought by the male to the young, does not increase with tail length (Møller 1992, 1994b).

Rather than direct benefits, female barn swallows choosing males with long tails seem to obtain indirect benefits—that is, they appear to secure superior genes for their offspring. This idea is supported by Møller's (1994b) demonstration of a positive correlation between offspring longevity and father's tail length. The effect is a strong one: young barn swallows increase in longevity by approximately 0.5 years for each increase of one standard deviation in father's tail length. Møller (1994b) attempted to eliminate the possibility that this effect was a direct, nongenetic one. He showed that male feeding rate did not increase with tail size, and furthermore that the father's tail length was not related to offspring mass (which often correlates with offspring survival). The young of males with long tails did not live longer because they were better fed as nestlings, which leaves good genes as the most likely explanation for their superior longevity.

Cross-fostering provides the best method for ruling out direct benefits as an explanation for the increased viability of the offspring of long-tailed males.

Møller (1990a) used this method to demonstrate that the offspring of long-tailed fathers inherit parasite resistance. Barn swallows are hosts to a number of parasites, including the blood-eating mite *Ornithonyssus bursa*, which has a variety of negative effects on the fitness of their hosts (Møller 1994a). Tail length is a good predictor of mite loads on males, the long-tailed males having fewer mites than males with short tails (Møller 1991b). This relationship suggests that a long tail signals that a male has genetic resistance to parasites, and cross-fostering experiments show that this resistance is passed to the offspring of long-tailed males. By exchanging half the chicks in each brood between nests, Møller (1990a) showed that the tail length of a male was strongly negatively correlated with the mite loads of his own (genetic) offspring, both when they were in his own nest and when they were in another nest. In contrast, neither a male's own mite load nor his tail length predicted the mite loads of unrelated offspring placed in his nest (Møller 1990a). These results together show that tail length is a reliable signal of a male's genetic resistance to parasites; this in turn may explain, at least in part, why male tail length is a reliable predictor of offspring longevity.

The potential indirect benefits obtained by mating with males having long tails have also been investigated in peafowl. Petrie (1994) captured eight feral peacocks and housed them in separate pens, each with four randomly chosen females. Offspring from the resulting matings were reared communally, apart from any of their parents. Offspring mass, measured at 84 days, was positively correlated with the father's train length, and with the size of the eye-spots in the train. In a multiple regression, eye-spot size was a better predictor of offspring mass than any other measure of the father's phenotype. Petrie (1994) released a sample of the young into the wild, and monitored their survival over the succeeding 2 years. Offspring survival over this period was significantly correlated with the father's eye-spot size. It is worth pointing out that in the peafowl work, a female preference was demonstrated for eye-spot number, whereas it is size of eye-spots that was shown to predict offspring size and survival. Various measures of the train may be intercorrelated, but the case for reliability would be more convincing if the receiver response and signal reliability were demonstrated for the same trait.

COSTS OF LONG TAILS

The huge trains of peacocks and the extremely long tails of long-tailed widowbirds seem likely to have substantial developmental costs, that is, costs in energy and nutrients invested in growing all that extra plumage. A comparative study of tail-feather morphology by Aparicio et al. (2003) revealed a negative relationship between sexual dimorphism in feather length and two measures of structural strength—the width of the rachis (the center support structure of the feather) and the number of feather barbs—among males but not females, suggesting that the developmental cost of longer feathers may be offset by making those feathers structurally more simple and thus less expensive to produce. Unfortunately, the developmental costs of elongated feather ornaments have not, to our knowledge, been investigated directly. Instead, most work on the costs of elongated tails has focused on the aerodynamic effects of tail length, especially effects on maneuverability and energy consumption.

Tails are different from many of the other signals we have discussed, in the sense that they have a natural selective advantage apart from any advantage in signaling. This selective advantage lies in enhancing flight performance, and because of this advantage all species of flying birds have a tail of some length, even if tail length is not used as a signal. This fact has major implications for attempts to measure the cost of tail elongation (Evans and Thomas 1997). Suppose that the handicap hypothesis is the hypothesis of interest, in other words the alternative hypothesis, or H_A. H_A then states that tail length is an indicator of male quality that is reliable because tail elongation is costly. A reasonable test might seem to be to elongate the tail farther and measure whether males pay increased costs, for example in decreased flight performance. Consider, however, what the null hypothesis ought to be: the H_0 for our H_A. A reasonable null hypothesis is that each male's tail is currently at the optimum length under natural selection for aerodynamic performance. Because this null hypothesis also predicts that further elongation will decrease flight performance, the elongation experiment does not discriminate between H_A and H_0 (Evans and Thomas 1997). What distinguishes the two hypotheses is that the handicap hypothesis proposes that the current tail lengths, without elongation, are already elongated past their natural selection optima, whereas the null hypothesis proposes that each tail is at its optimum. The crucial test therefore is to shorten the tail (Evans 1998). The handicap hypothesis predicts that shortening should decrease costs, that is, improve flight performance, at least until tail length is shortened all the way past its optimum. The null hypothesis predicts that any shortening should increase costs, that is, decrease performance. Thus, somewhat paradoxically, the crucial test for the costliness of tail elongation involves shortening the tail.

Aerodynamic theory predicts that the negative effects of elongation should depend on tail shape (Balmford et al. 1993). Negative effects on flight performance should be most severe for tails that are "graduated," such that the innermost feathers are longest, intermediate in severity for "pintails," in which only the innermost feathers are elongated, and least severe for "streamer" tails, where only the outermost feathers are elongated (figure 3.11). Evans and Hatchwell (1992) investigated the costs for a tail belonging to the intermediate category, the pintail of the scarlet-tufted malachite sunbird (*Nectarinia johnstoni*). They found that males with experimentally shortened tails increased their efficiency at catching insects in the air, relative to controls, and increased the amount of time they spent in flight. Both these results are compatible with

graduated

pintail

streamer

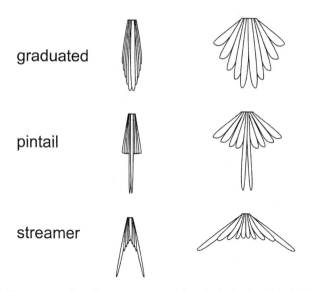

FIGURE 3.11. Three categories of exaggerated tails, from Balmford et al. (1993). The same tails are shown closed (left) and spread (right). Negative effects on flight performance are predicted to be greatest for graduated tails, intermediate for pintails, and least for streamer tails.

the prediction that the original, unmanipulated tail lengths of these birds are exaggerated past their aerodynamic optima.

More attention has been given to streamer tails, the last and least costly category of tail shapes. M. R. Evans (1998) shortened the tails of some male barn swallows and elongated the tails of others, each by 20 cm, and then filmed the birds in stereo while they were feeding nestlings. Evans found that shortening and elongation had the same effects on measures of flight performance; for example, both manipulations increased mean velocity and mean agility relative to controls. He concluded that both manipulations lowered flight performance, though he admitted that it is not obvious a priori that higher velocity and higher agility equate with lower performance. Even if we accept the measures of flight performance as valid, it is still possible that the aerodynamic optimum for tail length lies below the current average tail length, but less than 20 cm below it. Buchanan and Evans (2000) and Rowe et al. (2001) tested this possibility by reducing tail streamers in barn swallows in smaller increments, by 2, 4, 8, 10, 15, or 20 mm. Rowe et al. (2001) provided an intuitively satisfying index of flight performance by measuring the time it took individuals to negotiate an aerial maze, in which the swallows had to fly between rows of strings that narrowed down the length of the maze. Flight time was found to have a U-shaped relationship with manipulation, with the minimum flight time achieved at a manipulation of about 12 mm (figure 3.12). The important impli-

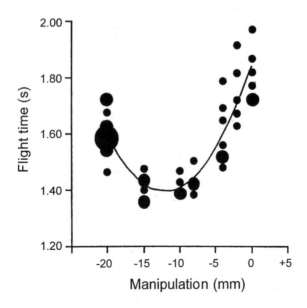

FIGURE 3.12. Flight times through an aerial maze for male barn swallows with manipulated tail lengths, from Rowe et al. (2001). Flight times are projected to be shortest for males with tails shortened by about 12 mm.

cation for signaling is that the natural tail lengths of barn swallows lie above the aerodynamic optimum; hence, their exaggeration beyond the optimum can be considered to be aerodynamically costly.

A problem with these aerodynamic studies is that measuring flight performance is a very indirect way to assess fitness. One can argue that a male with improved flight performance should be more adept at prey capture (Buchanan and Evans 2001), which might well lead to higher survival and reproduction; nevertheless, it would be more convincing to measure effects on survival and reproduction directly. Moreover, it is possible that streamers have one effect on foraging flight and another on other types of flight, such as the level, straight flight that must be most important during migration. Conceivably, then, shortening the streamers might depress some aspects of flight performance even as it improves others. By examining the effects of streamers on some more direct measure of fitness, such as survival or reproductive rate, one would hope to sum their overall effects on flight performance.

Møller (1989b) has examined the effects of tail manipulations on more comprehensive measures of fitness in barn swallows. Neither lengthening nor shortening tails had a measurable effect on the survival of males. Males whose tails were elongated in one year increased less in natural tail length between years than did controls, and perhaps as a consequence had more difficulty in attracting a mate in the second year. These results indicate that experimental elongation was costly to the males, but remember that this does not mean that their original tails were costly. Males with experimentally shortened tails did not experience greater tail growth than controls between years, nor did they experience better mating success in the second year. If the original tail lengths of these males had been costly, then shortening their tails should have improved these or other fitness measures (exclusive of their manipulation-year mating success); however, it is possible that the original tail lengths were above the optimum, and that the manipulation (20 mm) was great enough to move them back below the optimum. Thus the issue of whether the streamer tails of barn swallows are costly is not yet fully resolved.

Conclusions

We have concentrated on just three of the many signals that male animals produce to impress females: carotenoids in birds and fish, song in songbirds, and long tails in birds. Many other types of courtship signals have been investigated, of course, and some of these present aspects of great interest from the point of view of communication (Andersson 1994). Other notable examples include the courtship calls of frogs (Ryan 1980, Gerhardt 1994) and insects (Hedrick 1986, Brown et al. 1996). Frog calls in particular raise interesting issues, with regard to the importance of physical constraints in enforcing reliability; we will deal with these issues in chapter 4, in the context of male-male aggressive signaling.

The evidence for receiver response to courtship signals is unequivocal; females undoubtedly respond to intraspecific variation in male courtship signals. In some cases there is strong evidence that females respond to the signal in a way that augments male mating success, for example in some of the tail length and carotenoid studies. For some other mating signals, notably some of the song attributes, we have strong evidence of a behavioral response by females to the signal, but without clear evidence that the response enhances male mating success. These latter examples can be seen as lacking as demonstrations of sexual selection, but they provide all the evidence needed to establish the existence of a communication system, which is our main interest here.

The fact that females respond to mating signals does not necessarily imply that the signals are reliable indicators of male quality. Several alternative hypotheses predict that females will respond to courtship signals without assuming that females obtain useful information from those signals. Empirical assessment of the reliability of courtship signals is considerably complicated by the problem that we are never certain what it is that females wish to assess. The most we can conclude is that courtship signals do sometimes contain reliable information that ought to be of some value to females, such as information on heritable parasite resistance in the tail length of male barn swallows, or on heritable viability in the repertoire size of male great reed warblers.

All the courtship signals we have discussed are associated with plausible costs, but whether the costs are sufficient to enforce signal reliability is more open to question. Theory suggests that reliability is unlikely to be enforced by Zahavi handicaps, that is, handicaps that work as survival tests and that reveal male quality because only males of high quality can survive with the handicap. Much more workable in theory are condition-dependent handicaps, which function because only males in good condition can produce the display, and condition is tied to quality. Some of the mating signals we have discussed are clearly condition-dependent in a manner that fits well with theory. A good example is song output, which is subject to time and energy costs that are paid at the time that the signal is produced. Even if the direct energy cost of song is fairly negligible, as current evidence suggests, time devoted to the display must be taken away from other activities such as foraging, and some cost is thus inevitable as long as time is limiting for the signaler. Any signal with time and energy costs is a logical candidate for a signal of current condition, and evidence from provisioning experiments indicates that song output is indeed a reliable signal of current condition.

Tail length presents a more complex picture. The costs of a long tail are paid before, often substantially before, the moment at which the tail is displayed and a receiver responds to it. The past costs of a long tail are presumably of two kinds, first the developmental investment in producing the requisite mass of feathers, and second the cost of living with an impediment to aerodynamic efficiency. The latter cost, the only one that has been investigated, appears to be real. The difficulty is that the aerodynamic costs are paid after the tail is produced, and tail length can only be changed again at the next molt, which in general occurs after an interval of a year. The mechanism could work as follows: tail length in year $i - 1$ affects aerodynamic performance during that year, aerodynamic performance during year $i - 1$ affects condition at the next molt, and tail length for year i is adjusted to condition at the molt. Møller (1989b) presents evidence that adjustments of this type occur in barn swallows.

Some of the song features we have discussed are similar to tail length in imposing costs well before they are displayed to receivers; this is true, for example, of song complexity and local song structure. The cost structure of these features seems to be simpler than for tail length, however, in the sense that they presumably have only developmental costs, and no ongoing cost analogous to aerodynamic inefficiency. Once these features are set early in life, in many species there is no further opportunity to adjust them in response to changes in condition. Thus these song features in adulthood reveal, not an animal's current condition, but its condition during early development, potentially several years ago. The developmental-stress hypothesis gives an explanation for why past condition might still be relevant at the time of signaling, which is that condition in this early period is important, because that is when many other aspects of the phenotype develop and are set for life. Other species of birds do continue to modify their vocal behavior as adults, such as by adding new syllable types or song types to their repertoire, and birds are known to redevelop the brain nuclei responsible for song each year. In these cases, song might reflect a more recent condition state, for example condition in the previous winter, but even here it is not present condition being indicated, but condition sometime in the past.

As for carotenoid pigmentation, perhaps the best established cost is the increased risk of predation experienced by guppies and sticklebacks when they exhibit bright carotenoid colors. This cost is of just the type assumed by a Zahavi handicap, however, and thus probably should not be invoked to explain signal reliability. Moreover, there is little evidence that predation costs apply to carotenoid colors in birds such as house finches, and yet the signal seems as reliable in birds as in fishes. The investment of carotenoids in structures that produce outwardly visible colors seems to limit the use of these compounds in fighting parasites and disease in some species, and that cost may be important to reliability in those cases. Perhaps the simplest mechanism by which the reliability of carotenoid colors might be maintained is through their rarity in diets. Whether such rarity should be considered a cost or a constraint is a question we will return to later.

4 Signaling When Interests Oppose

When two unrelated animals compete for some resource, generally speaking one must win and the other lose—one will get the food, the mate, or the territory, and the other will not. A given outcome will benefit the winner and harm the loser, and in that sense the interests of the two are diametrically opposed. But it may be better for both contestants to settle the contest by signaling rather than by fighting, and therefore it is not surprising that a great deal of communication occurs in aggressive contexts. Questions of reliability and deceit seem particularly pressing in such contexts, for receivers should have no interest in attending to an opponent's signals of fighting ability or intentions unless those signals are honest. At the same time, deceiving a receiver might be particularly beneficial to a signaler in an aggressive context, given the absence of common interests.

The literature on aggressive signaling is again too vast for us to review comprehensively, so instead we have chosen particular systems to illustrate specific points of interest. First, we consider aggressive postural displays in birds, which were the focus of much of the early controversy on honesty in animal signals. Next, we examine "badges of status," a term that refers to plumage signals in birds that convey dominance status. Badges are intriguing because these features are simple and seemingly easy to produce, and yet appear to have a profound effect on fitness, raising the question of why deceit is not rampant. Third, we discuss weapon displays in decapod crustaceans, displays that exhibit the enlarged claws these animals use in fighting. Weapon displays in crustaceans are particularly interesting to us because they provide some of the best evidence for deception available for any animal signaling system. Finally, we review the role of the dominant frequency of frog calls in aggressive contests, a signal feature that some have argued is constrained to be honest because of its dependence on body size. Before we get to any of these signaling systems, however, we begin by reviewing theory relevant to signals of aggression.

Signaling in Aggressive Contexts: Theory

Suppose two animals compete for some resource. The two are evenly matched in fighting ability, but one values the resource more, and so is willing to fight harder to get it; we would say this animal is more aggressive. If a fight occurs, our more aggressive animal will win because of its greater willingness to fight, but both contestants will have to pay the cost of fighting. Both animals will therefore benefit if they honestly signal their level of aggressiveness at the start of the encounter, the less aggressive animal then conceding; the resource will be allocated in the same way as if there had been a fight, and neither animal will have to pay the cost of fighting.

The weakness of this naive argument for honesty is that the signaling system just pictured is highly vulnerable to cheating (Dawkins and Krebs 1978, Maynard Smith 1979). Cheaters will give a highly aggressive signal, no matter what their current motivation, and so will consistently win contests. Cheating thus is favored by selection and will spread through a population. Once cheating becomes sufficiently common, the signal is no longer informative, and there is no reason for receivers to respond to it. Once receivers cease to respond, a signaling system no longer exists. The same argument can be made against signals that honestly convey fighting ability rather than aggressive intentions.

An argument against the evolutionary stability of honest signals of fighting ability was formalized by Maynard Smith (1974), in the context of his "war of attrition" model. This model envisions a contest in which victory goes to the contestant willing to persist longer. Victory has a payoff, v, and the cost of the contest, m, is proportional to its duration. The first contestant chooses a particular cost, m_1, that it is prepared to pay, and wins if m_1 is greater than m_2, the cost chosen by the second contestant. Maynard Smith (1974) shows that no pure strategy of choosing a single m can be evolutionarily stable. Suppose that we have a population in which all individuals choose a cost m_1. Then if these individuals play each other they would each win half the time, giving them an expected payoff of $1/2v - m_1$. Another strategy, which chooses a cost m_2, could invade, as long as $m_2 > m_1$, because this second strategy would win every contest, giving a payoff of $v - m_1$. Clearly, the payoff to choosing m_2 is greater than the payoff to choosing m_1, so choosing m_1 is not an evolutionarily stable strategy (ESS). Maynard Smith (1974) shows that a mixed ESS does exist, in which contestants randomly choose m from a negative exponential distribution. Thus far, no signaling is involved, but Maynard Smith (1974) goes on to ask whether it can be advantageous for an individual to signal in advance its intended persistence, say through the intensity of its display. Maynard Smith's answer is no, because if the display is accepted by receivers, then the optimal strategy is to display at maximal intensity regardless of one's actual intentions. If all contestants give the maximal display, there is no information in the display, and receivers should again evolve to ignore it.

Honesty in aggressive signaling can be rescued by Zahavi's handicap principle. The first theoretical model to show explicitly how this rescue can be effected was provided by Enquist (1985); Grafen (1990b) labels this the first "model of biological signalling" of any kind. Because Enquist's model, like the Sir Philip Sidney game we introduced in chapter 1, is unusually transparent, we produce its logic in detail in box 4.1. The model assumes that contests occur between individuals that can be categorized as either strong or weak, and that contestants signal their strength with one of two displays, A or B. Honest signaling consists of producing A when strong and B when weak; a deceitful strategy is to produce A even if weak. The benefit of deceit is that a deceitful weak individual can successfully bluff another weak animal; the cost is that, by signaling strength, the weak individual may embroil itself in a fight with a strong one. Enquist (1985) assumes that the cost of fighting a stronger animal is greater than the cost of fighting an animal of equal strength. If the difference in these costs is great enough (relative to the value of the resource being contested), then honest signaling is an ESS. The requirements of the handicap principle are met, in the sense that the more effective signal is costly, and the cost falls more heavily on individuals of lower quality. Note that exactly the same logic applies if the signals are taken to communicate aggressive motivation rather than strength; thus Enquist's model shows that such signals, whether of fighting ability or motivation, can be reliable.

Grafen's (1990b) general signaling model confirms that relative fighting ability can be communicated reliably in aggressive contests. We discussed Grafen's model earlier in the context of males signaling their quality to females (chapter 3), but Grafen (1990b) shows that the model's logic also applies to males signaling fighting ability to each other. The model portrays a situation in which one individual signals and another assesses, rather than mutual assessment as in Enquist's model. Assume that males vary in fighting ability and that they produce a display reflecting that ability, say an exhausting display that reflects endurance. Assume that the receiver uses the display to assess the signaler's fighting ability, that the receiver benefits from making an accurate assessment (i.e., either overestimating or underestimating one's opponent is costly), and that the signaler benefits from a more positive assessment by the receiver. Then if males of high fighting ability benefit at least as much from increases in display as do males of low ability, and if the cost of display increases more rapidly with display intensity in males of low ability, then reliable signaling is evolutionarily stable.

These and other models show that, if certain assumptions are met, honest communication of fighting ability or aggressive intentions can occur in contest situations. The specified assumptions seem more probable than not. So what about cheating? Is deceit possible in aggressive signaling? Note that if the assumptions of the above two models fail, it does not necessarily follow that deceitful signals will occur; rather, at equilibrium there may be no signaling at all. We need to know whether there can be an evolutionarily stable signaling system in which deceitful signals of aggressive intentions or fighting ability play a role.

Box 4.1.
SIGNALING THROUGH CHOICE OF BEHAVIOR: ENQUIST'S (1985) MODEL 1

In a population consisting of equal numbers of strong and weak individuals, two individuals contest for some resource. Each is assumed to know its own strength, but not the strength of its opponent. In the first step of the contest, the contestants choose to produce either of two displays, A or B, where A signifies a strong individual and B a weak. In the second step, the contestants may give up, attack, or produce display A again and then attack if the opponent does not concede.

Enquist (1985) seeks to determine whether a strategy of honest signaling can be evolutionarily stable. The honest-signaling strategy obeys the following rules: If strong, produce A in step 1. In step 2: if opponent produces A, attack; if opponent produces B, repeat A and attack if opponent does not concede. If weak, produce B in step 1. In step 2: if opponent produces A, give up; if opponent produces B, attack.

Enquist (1985) then determines whether the honest-signaling strategy is proof against invasion by a bluffing strategy, in which weak individuals produce display A. Let v be the value of victory, c be the cost of a fight between equals, and d the cost to a weak individual of being attacked by a strong. If you are weak and honest and encounter a *weak* opponent, then your average payoff is $1/2v - c$ (assuming you have a 50:50 chance of winning the resource, which is worth v, at the cost of a fight between equals, worth $- c$). If you are weak and honest and encounter a *strong* opponent, your average payoff is 0 (assuming you never win the resource and never have to pay a cost for fighting). Remembering that weak and strong opponents are equally abundant, the overall average payoff for the honest weak individual is $1/2(1/2v - c)$.

If you are weak and dishonest, and encounter a *weak* opponent, your payoff is v (assuming that the dishonest signaler can always successfully bluff a weak opponent without a fight, and thus will always win the resource (+v) at no cost). If you are weak and dishonest and encounter a *strong* opponent, your payoff is $- d$ (assuming that the dishonest signaler cannot win against a strong opponent and must pay the cost $(- d)$ to a weak individual of being attacked by a strong one). Again assuming that strong and weak individuals are encountered at equal frequencies, the overall average payoff for the weak, dishonest individual is thus: $1/2v - 1/2d$.

It follows, then, that honesty is the best policy if

$$1/2(1/2v - c) > 1/2v - 1/2d, \text{ or}$$
$$1/2v - c > v - d, \text{ or}$$
$$d - c > 1/2v$$

This inequality can be satisfied as long as the cost to a weak individual of being attacked by a strong opponent (d) is sufficiently greater than the cost of fighting another weak individual (c). Thus the model does not say that honest signaling *must be* evolutionarily stable, but rather that it *can be* evolutionarily stable.

Abstracted with permission from M. Enquist. 1985. Communication during aggressive interactions with particular reference to variation in choice of behavior. *Animal Behaviour* 33:1152–1161. Elsevier. ∎

Grafen (1990b) suggests that deceit can occur in his model if a second set of males exists for whom signaling is cheaper than for the original set. The optimal level of display would be higher for a male in this second set than for a male of equal fighting ability in the original set. To a receiver used to the original set it would seem that males from the second set were deceitfully exaggerating their ability. Why the necessary differences in signaling costs would occur is not obvious, however.

Adams and Mesterton-Gibbons (1995) offer a model that explicitly shows how honest and deceitful signals can coexist in a stable system. Here, deceit is possible because receivers pay a cost if they probe for deceit, as suggested by Dawkins and Guilford (1991). The model considers a situation in which one animal defends a resource against another. The defender can either threaten the challenger or defend without threatening, and the challenger will then decide whether to attack or concede. Alternative versions of the model give the threat display either a production cost (such as energy), which is always paid, or a vulnerability cost (in terms of increased risk of injury), which is paid only if the opponent attacks and beats the signaler. Vulnerability costs are imposed in addition to the normal costs of fighting. Only the version incorporating a vulnerability cost yields a stable signaling system.

Two types of signaling systems are possible at equilibrium. In one, all defenders above a certain threshold in fighting ability produce the threat, and all those below the threshold do not; the signal is then, in a sense, completely honest. The second possibility is for two thresholds to exist; call them I and J, with I < J. Defenders will threaten if their fighting ability is either greater than J (the upper threshold) or below I (the lower threshold); only those of intermediate strength (between I and J) do not threaten. The production of the threat by those weakest in fighting ability is a nonintuitive outcome of this model, and can be regarded as a form or bluffing or deceit.

Why is deceit possible in the Adams and Mesterton-Gibbons (1995) model, and not in models such as Grafen's (1990a)? Adams and Mesterton-Gibbons assume that vulnerability costs are imposed only if the signaler provokes and then loses a contest, which means these costs are higher for weak individuals, who are more likely to lose (figure 4.1). Display is thus costlier for weak individuals than for strong, which is in accord with the assumptions of other handicap models, such as Grafen's. What differs in the Adams and Mesterton-Gibbons model is that the benefits of display are also greater for weak individuals than for strong. One benefit of producing the threat is that threatening enables the defender to deter some proportion of challengers, and in this way retain the resource. Strong defenders would be able to beat many of these challengers anyway, so this benefit is greater for weak individuals (figure 4.1). Put another way, for a very weak defender a successful bluff may be the only possible method of winning, whereas a very strong defender will win regardless. A second benefit of display is that a successful threat enables the defender to avoid

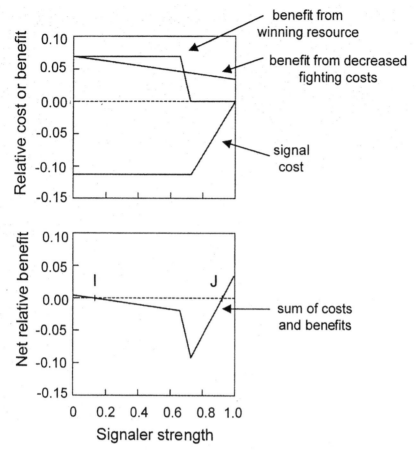

FIGURE 4.1. The costs and benefits of aggressive signaling in a model that predicts bluffing, from Adams and Mesterton-Gibbons (1995). Signaler strength is measured as the proportion of other individuals the signaler can defeat. The aggressive signal has two kinds of benefits (upper panel). The benefit from winning the resource is the product of the value of the resource times the increased likelihood of winning due to signaling. Only receivers below a certain threshold strength can be deterred by threats; signalers above that threshold experience no increased likelihood of winning, and so for them this benefit is 0. The second benefit of signaling is that the costs of fighting are saved if the opponent can be induced to give up without a fight. Fighting costs go down as a linear function of an individual's strength, and consequently so does the benefit from decreased fighting costs. The signal cost (upper panel) is due to an increased vulnerability to injury if the receiver attacks and the signaler loses; consequently, this cost is low for the strongest individuals, who are unlikely to lose. The sum of the costs and benefits of signaling (lower panel) is positive for the strongest individuals (with strength > J), but is also positive for the weakest individuals (with strength < I). The lines in both panels illustrate a solution for a specific resource value, vulnerability cost, and fighting cost/strength relationship.

a possible fight. Fighting costs are assumed to be higher for weak individuals than for strong, because the weak individuals are more likely to lose. If the summed benefits of display are great enough for weak individuals, they can outweigh the high costs, making threats the optimum behavior for the weakest individuals. This kind of solution, with a mix of honesty and bluffing, becomes more likely as fighting costs increase, and as vulnerability costs decrease.

Számadó (2000) provides a second model of aggressive signaling that allows some level of dishonesty. Számadó (2000) bases his analysis on Enquist's (1985) model (box 4.1), which assumes that assessment is mutual, rather than one-way as it is in the model of Adams and Mesterton-Gibbons (1995). Számadó (2000) shows that, given the right parameter values, a mixed signaling strategy can be an ESS in a contest with mutual assessment. In this mixed strategy, strong individuals behave as in Enquist's original model, giving the signal of strength A, and then attacking if the opponent does not concede. Weak individuals give either the weak signal B, with probability $1 - p$, or the strong signal A, with probability p. A weak individual that gives B withdraws if the opponent signals A but attacks if the opponent signals B; a weak individual that gives A concedes if the opponent produces A but waits until the opponent withdraws if the opponent signals B. The model depends on several key assumptions. First, the cost to a weak individual of a fight with a strong one (C_{ws}) is greater than the cost to a strong individual of fighting with another strong individual (C_{ss}). Second, the cost to a weak individual of fighting with another weak individual (C_{ww}) is greater than the cost to a strong individual of fighting with a weak one (C_{sw}). Finally, the average payoff from a fight with an individual of equal strength is greater than the cost ($1/2v > C_{ww}$ and $1/2v > C_{ss}$). Given these assumptions, and depending on parameter values and on the proportion of weak individuals in the population, cheating can occur at high frequencies at a stable equilibrium.

The mixed strategy of reliable and deceitful signaling allowed in the Számadó (2000) model means either that each weak individual signals dishonestly on some proportion p of occasions and honestly the rest, or that a proportion p of weak individuals always signal dishonestly while the rest always signal honestly. In either case, no difference exists between the weak individuals that signal honestly and those that signal dishonestly. By contrast, in the Adams and Mesterton-Gibbons (1995) model, fighting ability is continuous, and it is only the weakest of the weak that bluff. Consequently, there is a difference between weak individuals that signal honestly and those that signal dishonestly (the latter being even weaker). In both models, some proportion of those individuals giving the signal of strength actually are strong. This proportion does not necessarily have to be greater than half (Számadó 2000), but it does have to be high enough that responding to the signal of strength as if the signaler is strong continues to be advantageous to receivers.

The later models, then, allow for both reliability and deceit in aggressive signals. Deceit is made possible by reliability, because without reliability no one would respond to the signal. To enable sufficient reliability to exist, the handicap assumptions must be met: the most effective signals must be costly, and the cost must fall disproportionately on individuals of poor quality.

Postural Displays of Aggression in Birds

Aggressive displays in birds have figured prominently in the debates over the reliability of aggressive signals. Early empirical studies attempted to show that avian aggressive displays both predict the subsequent behavior of the signaler and affect the subsequent behavior of the receiver (Stokes 1962, Dunham 1966, Andersson 1976). With the advent of game theory, Caryl (1979) argued that the then-existing models predicted that aggressive displays should be neither reliable nor effective. Caryl reinterpreted the results of some of the earlier empirical studies, in an attempt to reconcile their implications with the then-current game-theory predictions. In particular, Caryl argued that displays were not good predictors of attack, since the displays that best predicted attack were actually followed by attack (on the next move) only about half the time. This argument implicitly assumes that immediate intentions are what these displays have evolved to communicate.

Caryl (1979) makes his argument against reliability in the context of Maynard Smith's (1974) war of attrition model, discussed earlier. In that model, what determines contest outcome is persistence, and thus the important information that might or might not be communicated is the sender's intended persistence time. Clearly, one could signal honestly about persistence time without revealing one's next move in a contest, and the data on predicting attack are thus not that germane to the original model. Caryl (1979) generalizes the war of attrition by assuming that what determines contest outcome is the maximum cost that a contestant is willing to pay rather than persistence time, but this does not change the argument substantively: unless a single attack contributes an overwhelming proportion of the total cost of the contest, one can signal reliably about total cost without signaling reliably about attack-on-the-next-move.

We can derive more general predictions about signaling the likelihood of attack-on-the-next-move from Enquist's (1985) model, which explicitly includes immediate intentions as part of the modeled strategies. Remember that in this model (box 4.1), the ESS was:

If *strong*, produce A in step 1. In step 2: if opponent produces A, attack; if opponent produces B, repeat A and attack if opponent does not concede.

If *weak*, produce B in step 1. In step 2: if opponent produces A, give up; if opponent produces B, attack.

Note that in this strategy the signal of strength or aggressive motivation, A, is not always followed by attack. In fact, if the above strategy prevails in the population, and strong and weak individuals are equally common, then a focal individual that signals A will follow up with attack only 50% of the time (when the opponent is also strong). Thus, in a completely honest signaling system, the more aggressive signals predict immediate attack only half the time. The signals are honest about overall aggressive motivation or fighting ability, not about the next move in the contest.

The point we are making is not that theory rules out reliable signaling of immediate intentions. Rather, it is that a signal can be reliable about something that is of interest to an opponent in an aggressive contest without being reliable about immediate intentions. With that caveat in mind, we can turn to the data on postural displays in birds.

RECEIVER RESPONSE TO AVIAN POSTURAL DISPLAYS

For birds, the typical method for determining whether receivers respond to postural displays has been to observe natural aggressive encounters between pairs of birds and correlate a signaler's display with the receiver's next behavior (Stokes 1962, Dunham 1966, Andersson 1976, Waas 1991, Scott and Deag 1998). Results of this method should be viewed with caution, however, because the receiver's behavior may be due to some factor that covaries with the signaler's display, rather than to the display itself. Only by manipulating the signaler's display experimentally could this alternative interpretation be ruled out.

Stokes' (1962) classic study of blue tits (*Parus caeruleus*) interacting at a winter feeder provides a representative data set. Stokes lumps receiver response into three categories: attack, escape, or stay. Cases in which the signaler attacks are eliminated, to increase the likelihood that any change in receiver behavior is due to the signaler's display rather than to other aspects of its behavior. Stokes uses simple contingency analysis to test whether the behavior of the receiver differs depending on whether one of eight displays is or is not given by the signaler, and finds highly significant differences for seven of the eight displays. For example, if the signaler gives a "crest erect" display, the probability that the receiver escapes decreases from 37% to 3%; conversely, if the signaler gives a "wings raised" posture, the probability that the receiver escapes increases from 27% to 48%. Changes in receiver response are not always so substantial in other studies, but in general this kind of analysis supports the conclusion that birds change their behavior in response to the aggressive displays given by other birds.

RELIABILITY OF AVIAN POSTURAL DISPLAYS

The reliability of avian postural displays has been assessed in the same way as their effectiveness—by observing natural encounters and correlating the display given with the next behavior, in this case the next behavior of the signaler. One problem with this method, as we discussed above, is that the display might carry reliable information about fighting ability or overall aggressive motivation without being accurate about the next move; nevertheless, it is interesting to ask if any information about the next move is in fact revealed.

We can again start with Stokes' (1962) study of blue tits. Caryl (1979) uses the results of this study to argue that blue tit displays are relatively uninformative about attack likelihood. More specifically, he argues that blue tit displays are less informative about attack than about escape; the combination of postures that best predicts attack is followed by attack only 48% of the time, whereas the combination that best predicts escape is followed by escape 94% of the time. This comparison, however, is not the most relevant one because escapes occur overall just over twice as often as attacks in Stokes' data (even though attack is defined broadly; see Scott and Deag 1998). Thus even if the association between display and behavior were random, we would expect any given display to be followed more often by escape than attack. What is more pertinent is a contingency analysis, asking whether attack is more likely following a particular display than in the absence of that display, or following one display rather than another. Stokes found that blue tits attacked after 40% of 225 instances in which they gave the "body horizontal" display compared to 7% of 457 instances in which they did not give this display; our analysis shows the association of the display with attack to be highly significant (G statistic = 105.9, P < 0.0001). Strong relationships between other displays and particular behaviors are also evident in Stokes' data. These relationships demonstrate that the displays provide information that is at least somewhat reliable on what the signaler will do next. Other studies showing relationships between avian postural displays and signaler's next behavior include Andersson (1976) on great skuas (*Catharacta skua*) and Waas (1991) on little blue penguins (*Eudyptula minor*).

In aggressive interactions, we might expect the behavior of the signaler to be contingent on the behavior of the receiver, as well as on its own prior display (van Rhijn 1980). This kind of contingency is included in Enquist's (1985) model, wherein the signaler follows the display of strength with an attack only if the receiver does not retreat first. If signaler behavior is contingent on receiver response in this way, we cannot expect to predict the signaler's behavior from its signals unless we control for receiver response. Caryl (1979) recognized this problem, but downplayed its importance.

The right kind of statistical analysis allows one to control receiver response while looking at signaler's behavior (Nelson 1984). Such an analysis is illustrated by Popp's (1987) study of American goldfinches. Popp videotaped aggressive encounters at a winter feeder to determine sequences of signaler's display, receiver's response, and signaler's next act. Goldfinches use three main displays in aggressive contexts: low-intensity head forward (LHF), high-intensity head forward (HHF), and wingflap (WF). LHF is least effective and

WF most effective in eliciting retreat from receivers, and HHF is intermediate (figure 4.2a). In parallel with that result, signalers attack following LHF least often and following WF most often, and HHF is again intermediate (figure 4.2b). Popp used log-linear models to show that the signaler's display type is strongly related to the receiver's response, and that the receiver's response is strongly related to the signaler's next act. Controlling for these relationships, the signaler's display was still strongly predictive of the signaler's next act. Thus, displays contain information about signaler's next behavior with receiver's response controlled.

Popp (1987) also provided an experimental test of signal reliability. From the acts that follow the goldfinches' three displays, we can infer that the displays are graded signals of aggression, with LHF < HHF < WF. Popp manipulated aggressive motivation by depriving captive goldfinches of food, and then observing how display use changed during competition for food. Food-deprived birds showed significant changes, decreasing their use of the least-aggressive LHF display and increasing their use of the most-aggressive WF display. The proportion of encounters ending in fight also increased with food deprivation. Display use changed when aggressive motivation was manipulated, confirming that the displays are at least somewhat reliable signals of aggressive motivation.

Costs of Avian Postural Displays

Avian postural displays are unlikely to have significant production costs, given the rather slight movements that usually are involved. Instead, costs, if they exist at all, seem likely to be receiver-dependent. Such receiver-dependent costs can be incurred in two ways: (1) the more effective displays may be more likely to elicit attack from an opponent, and (2) the more effective displays may make the signaler more vulnerable to injury, if an attack should occur. These can be termed retaliation costs and vulnerability costs, respectively (Vehrencamp 2000).

Models of honest signaling in aggressive contests often incorporate retaliation costs to ensure reliability. In Enquist's (1985) model, for example, the signal of strength is costly because it invites an attack from strong individuals; this cost falls more heavily on weak individuals because the cost to a weak animal of fighting a strong opponent is greater than the cost to a strong individual of fighting another strong individual. Zahavi (1987), however, emphasized vulnerability costs in his verbal arguments for reliability. He suggested that threat displays often involve assuming a posture that makes warding off attack particularly difficult. An animal thus may threaten by exposing a vulnerable body part, by advancing within the opponent's striking range, or by relaxing its muscles to show that it is not braced for an attack.

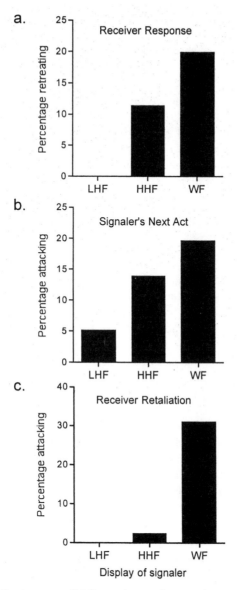

FIGURE 4.2. The effectiveness, reliability, and cost of aggressive postures in American goldfinches, from Popp (1987). Three display postures are examined: low-intensity head forward (LHF), high-intensity head forward (HHF), and wingflap (WF). **a.** The three displays can be ranked on their effectiveness in causing the receiver to retreat as LHF < HHF < WF. **b.** The ranking on aggressiveness, measured as the likelihood that the signaler's display will be followed by an attack, is the same as for effectiveness: LHF < HHF < WF. **c.** The most effective and aggressive displays are also the most likely to elicit an attack from the receiver.

Empirical studies have concentrated on retaliation costs, measured as increased risk of being attacked, rather than vulnerability costs, which would have to be measured as increased risk of injury per attack. Popp (1987), for example, assessed risk of being attacked in his study of American goldfinches. Remember that the three main aggressive displays given by goldfinches showed the same ranking in effectiveness in causing opponents to retreat as they did in their reliability in predicting signaler attack (LHF < HHF < WF in both cases). Popp also found that this same ranking applied to the probability of eliciting an attack from the opponent (figure 4.2c). In other words, the signals that are most effective in causing the opponent to retreat are also the ones most likely to provoke the opponent to attack. This pattern may seem paradoxical, but it is exactly what is necessary if retaliation costs are to ensure reliability, because the most effective signals should be the most costly.

Waas (1991) considered both retaliation and vulnerability costs in his study of little blue penguins. The 22 displays given by these penguins were placed in five categories: distance increasing (S1), defensive stationary (S2), offensive stationary (A1), distance reducing (A2), and contact (A3). Of the three categories considered aggressive, two (A2 and A3) were effective, in the sense of significantly increasing the probability that the opponent would retreat. Only one of these display categories (A2) was a significant predictor of attack by the signaler. None of the aggressive displays significantly increased the chances that the opponent would attack, and in fact opponents were in general more likely to attack following submissive displays (S1 and S2) than following aggressive ones. There is thus not much evidence that effective displays carry retaliation costs in this system, but Waas (1991) argues that they do carry vulnerability costs. Aggressive displays of little blue penguins involve orienting toward the opponent, thus exposing to attack such vulnerable areas as the eyes and throat, whereas all submissive displays involve turning away from the opponent, which hides these areas. The contention that aggressive displays increase vulnerability is thus plausible, but no quantitative evidence exists to support the argument.

As we have argued for other signaling systems, more needs to be done in measuring the costs of avian postural displays. Attack and vulnerability costs could be combined, for example, using injuries as a common currency. Because injuries are rare events in fights between birds, a very large sample size would be needed in such a study. Moreover, because injuries can differ in their severity and hence in their ultimate fitness costs, it might be necessary to take severity of injury into account in measuring costs. Despite the difficulties, studies combining attack and vulnerability costs are needed if we are to determine whether more effective displays are more costly, as predicted by theory.

Conclusions

We do not know whether postural displays are used deceitfully in aggressive encounters between birds, and it seems likely that our ignorance on this point will continue. One reason is that the reliability of these displays is usually measured by whether the displays predict the signaler's next behavior, which after all may not be what the displays have evolved to communicate. A second reason is that the signaler's next act may be contingent on factors, such as the receiver's response, that can change between the point at which the display is produced and the point at which the next act is performed. Therefore, if a display that statistically predicts attack is not followed by attack in a specific instance, we cannot know whether the signaler (a) was being deceitful, (b) was being honest about something other than its next move, or (c) was being honest about its next move but then some contingency intervened to change that move.

We are left with many unanswered questions concerning avian postural displays. What we can conclude is that such displays are at least somewhat effective in influencing the behavior of opponents, and that they are at least somewhat reliable about the signaler's next move. What we cannot conclude is whether avian postural displays have costs of a nature that could enforce their reliability, and whether these displays are ever used deceptively.

Badges of Status

In winter, Harris's sparrows (*Zonotrichia querula*) vary greatly in the amount of black plumage they exhibit on their face and throat. Males in general are blacker than females, and adults are blacker than first-year birds, but there is considerable variation within as well as between age-sex classes (Rohwer et al. 1981). Rohwer (1975) proposed that the amount of black serves as a status signal, with darker plumage signaling dominance and lighter plumage signaling subordinance. As evidence, he showed that in aggressive interactions between two sparrows, the darker bird won substantially more than half the time. Subsequently, other species of birds were also found to have variable plumage traits that correlate with dominance. These plumage signals are long lasting, and can be altered (for the most part) only when the feathers are molted, which usually happens just once or twice a year. The signals also appear cheap to produce, in the sense that they require only the manufacture and deposition of small amounts of seemingly inexpensive pigment. This type of long-lasting, cheap-to-produce signal of dominance has been termed a "badge of status" (Dawkins and Krebs 1978).

Several models have investigated the evolutionary stability of status signaling. Maynard Smith and Harper (1988) assumed that, in a status-signaling system, contests are settled by relative aggressiveness, defined as the willingness to fight longer or harder. Assuming that aggressiveness has no cost independent of fighting, they asked whether a stable polymorphism in aggressiveness is possible. Their model assumes that the fitness benefit of aggressiveness to an individual of aggressiveness m is bz, where b is the benefit of winning a contest and z is the proportion of the population whose aggressiveness is less than m and which our focal individual therefore can defeat. This proportion increases as m increases, according a benefit to high m. Aggressiveness has a cost, because our focal individual will have to fight when it encounters another individual of equal aggressiveness, which happens with probability p(m). The cost of such a fight, C(m), is assumed to increase as aggressiveness increases. The net fitness of an individual of aggressiveness m can therefore be written as bz − p(m)C(m). Intuitively, one can see that if the cost increases rapidly enough, this rising cost can balance the increase in probability of winning with increasing m, so that all levels of m lead to equal fitness, and a stable polymorphism is possible.

Given a stable distribution of levels of aggressiveness, Maynard Smith and Harper (1988) ask whether signals that honestly reveal aggressiveness are possible. To answer, they consider the fate of a bluffing phenotype, one that signals a level of aggressiveness, m_2, higher than its actual level, m_1. The bluffer is assumed to be able to defeat by bluffing any honest individual of aggressiveness between m_1 and m_2. The overall success of bluffing versus honesty depends on what happens to the bluffer when it confronts an honest signaler showing m_2. If, on the one hand, the bluffer can escape such a contest without paying the appropriate fighting cost, then its fitness is higher than that of an honest signaler whose aggressiveness is m_1; the bluffer wins some contests that the honest signaler would lose, and never has to pay extra costs. Bluffing is then advantageous and honest signaling is not stable. If, on the other hand, the bluffer has to pay the high cost of fighting honest m_2 individuals, then it has lower fitness than an honest m_2: it loses contests with honest m_2's and still has to pay their level of cost. All honest individuals have the same net fitness, so if our bluffer has lower fitness than honest m_2's it has lower fitness than all honest phenotypes. Because bluffing is then disadvantageous, bluffers cannot invade the honest signaling system.

An honest signaling system can, however, be invaded by another dishonest phenotype, one that signals it is less aggressive than it really is. Johnstone and Norris (1993) investigate the fate of this "modest" phenotype in a discrete signaling system, in which only two levels of aggressiveness can be signaled. They show that "modest liars" can invade and destabilize an honest signaling system unless there is a cost of aggressiveness that every aggressive individual experiences, regardless of whether it engages in fights. Johnstone and Norris suggest that such a cost might arise from the effects of testosterone on traits such as immune function and metabolic rate.

Hurd (1997) has pointed out that the status-signaling models of Maynard Smith and Harper (1988), Johnstone and Norris (1993), and others are all predicated on the assumption that subordinates and dominants have equal fitness, a balance maintained via frequency dependence. Subordinates are assumed to have the same basic fighting ability as dominants, and the latter win contests only because they are willing to fight longer and harder. If dominants and subordinates are of equal fighting ability, then it is necessary to assume that their differing levels of aggressiveness produce equal fitness, in order to explain why variation in aggressiveness persists. For the balance to be stable, fitness has to be frequency-dependent. The assumption of equal ability seems inappropriate, however, for the many status-signaling systems in which individuals with large badges and high status tend to be old rather than young, male rather than female, and/or large rather than small (Rohwer 1975, Rohwer et al. 1981, Balph et al. 1979, Fugle et al. 1984, Veiga 1993, Pärt and Qvarnström 1997).

Abandoning the assumption of equal abilities, Hurd (1997) offers a model of status signaling based on Enquist's (1985) aggressive-signaling model (box 4.1). As before, individuals can be either strong or weak in fighting ability and can give either of two signals: A denoting strength and B denoting weakness. Because signals have no production costs, signaling costs are exclusively receiver-dependent, incurred through fighting. The cost of a fight depends on the difference in fighting ability between two opponents. Thus the cost to a weak individual of fighting a strong one is $C(1)$; the cost of fighting an opponent of equal ability is $C(0)$; and the cost to a strong individual of fighting a weak one is $C(-1)$, with $C(1) > C(0) > C(-1)$. The honest-signaling strategy is the same as in the earlier model: give signal A when strong and B when weak; attack opponents that signal equal strength; flee from opponents who signal greater strength; and pause, then attack, for opponents signaling lower strength.

Where Enquist (1985) analyzed the stability of the honest-signaling strategy only against bluffing (signal strength, whether strong or weak), Hurd (1997) considers two additional alternatives. Remember that what disrupts honesty in the Johnstone and Norris (1993) model is the "modest" phenotype, which is aggressive but has a small badge. The two new alternatives that Hurd considers are variations on this modesty theme, and are appropriate for a case in which the signal reveals strength rather than aggressiveness. The "Trojan sparrow" (after Owens and Hartley 1991) behaves as an honest signaler if weak, but if strong it signals weakness and then attacks. Because the Trojan sparrow, when strong, has to attack weak, honest individuals that would have fled from an honest signaler, this strategy is disadvantageous as long as $C(-1) > 0$, which we can assume will always be true. The second modest strategy is illustrated by the "coward," which always signals weakness and always flees. Even when strong, the coward avoids fights with other strong individuals, and naturally fails ever to win the resource. Consequently, the coward strategy can invade the honest-signaling system only when the cost of fights between equals is considerably greater than the payoff of victory. This criterion places a further constraint on honesty, beyond that already imposed to ensure stability against bluffing (box 4.1), but a considerable range of parameter values still allows honest signaling as a stable strategy (Hurd 1997).

There are thus two genres of status-signaling models, one in which dominants and subordinates are of equal ability and their badges convey aggressiveness, and another in which dominants and subordinates are of unequal ability and their badges convey that distinction. In our opinion, the latter set of models is more relevant to the analysis of status signaling in birds, given the existence of phenotypic correlates of badge size such as age, sex, and body size.

RESPONSE TO BADGES

Suppose that we find that badge size is correlated with success in aggressive encounters in a given species. One explanation for the correlation is that badge size is used as a signal of fighting ability: a correlation exists between badge size and fighting ability, receivers attend to badge size and are intimidated by large ones, and large badges therefore help signalers win contests. A second possible explanation assumes no signaling via badge size: badge size is correlated with fighting ability, receivers nonetheless ignore badge size, and individuals with large badges win encounters simply because of their superior fighting ability. Because both the signaling and the nonsignaling hypotheses predict the original correlation between badge size and winning, this correlation by itself does not test the signaling hypothesis. Instead, it is necessary to manipulate badge size to test whether signaling occurs; if success in aggressive encounters changes following the manipulation, then badge size must have some direct influence on receivers.

Experimental manipulation of badges has been accomplished in a number of studies of birds in flocks, but only a subset of these have used sample sizes large enough to make the results convincing. One example is the study of white-crowned sparrows by Fugle et al. (1984). These birds vary in the boldness of their crown plumage: at the low end of the scale they have drab brown and tan stripes, and at the high end they have strongly contrasting black and white stripes. Adults tend to have bolder stripes than first-year birds, and males tend to have bolder stripes than females. In one set of trials, Fugle et al. (1984) used five captive groups of first-year females, each group with eight to ten birds. The crown plumage of some of these birds was painted with black and white enamels to resemble the crowns of adults. The remainder (the controls) were painted with either clear enamel or brown and tan enamels, so that the appearance of their badges would be unchanged. After treatment, the birds were placed together in aviaries, and relative dominance was determined from the outcome of aggressive encounters. The treatments had a very strong effect

Subadults:

Group 1: $E_R \rightarrow E_P \rightarrow E_W \rightarrow C_R \rightarrow C_W \rightarrow C_B \rightarrow C_K \rightarrow C_P$

Group 2: $E_B \rightarrow E_K \rightarrow E_W \rightarrow E_R \rightarrow E_P \rightarrow C_R \rightarrow C_K \rightarrow C_B$

Group 3: $E_N \rightarrow C_R \rightarrow E_Z \rightarrow E_P \rightarrow E_G \rightarrow E_B \rightarrow C_K \rightarrow C_Y$

Group 4: $E_Z \rightarrow C_G \rightarrow E_R \rightarrow E_K \rightarrow E_W \rightarrow E_B \rightarrow C_P \rightarrow C_S \rightarrow C_Y$

Group 5: $E_R \rightarrow E_L \rightarrow E_O \rightarrow E_B \rightarrow E_S \rightarrow E_Y \rightarrow C_E \rightarrow C_P \rightarrow C_D \rightarrow C_K$

Adults:

Group 6: $E_P \rightarrow E_I \rightarrow E_K \rightarrow E_D \rightarrow C_E \rightarrow C_L \rightarrow C_R \rightarrow C_W \rightarrow E_S \rightarrow C_Y$

Group 7: $E_P \rightarrow E_K \rightarrow E_R \rightarrow E_B \rightarrow C_W \rightarrow C_D \rightarrow C_E \rightarrow C_Y \rightarrow E_I \rightarrow C_L$

Group 8: $E_B \rightarrow E_W \rightarrow E_K \rightarrow E_Y \rightarrow E_P \rightarrow C_L \rightarrow C_R \rightarrow C_E \rightarrow C_O \rightarrow C_D \rightarrow C_I$

FIGURE 4.3. The effect of experimental manipulation of a badge of status on dominance in female white-crowned sparrows, from Fugle et al. (1984). E indicates an experimental bird; C indicates a control. The crowns of the randomly chosen experimental birds were painted a bold black and white, while the crowns of controls were painted brown and white in subadults and black and gray in adults. We depict the hierarchies as linear, though a few triangular relationships were observed.

on dominance. In three of the five groups, every badge-enhanced bird was dominant to every control bird; in the other two groups, the badge-enhanced birds dominated most, but not all, of the controls (figure 4.3). Overall, 23 of 24 experimental badge-enhanced birds had winning records against controls, a highly significant result. A second set of trials with three groups of adult females had similar results: birds painted to have bolder crowns ended up with higher dominance ranks (figure 4.3).

Rohwer (1985) manipulated badges in Harris's sparrows, the species in which status signals were first described. He performed two trials, one with 15 first-year males and the other with 12 first-year females. In both trials, approximately half the birds were treated with black hair dye to produce badges like those of adult males. Controls were treated only with shampoo and water, leaving them with their original, smaller badges. Among the first-year males, all seven dyed birds were dominant to all eight controls. Similarly, among the first-year females, all six dyed birds were dominant to all six controls. Overall, dyed males won 283 of 286 interactions against control males, and dyed females won 164 of 164 interactions against control females.

In a sense, the above two experiments exaggerate the impact of badges on dominance, for in both studies the researchers took care to hold constant other variables that might affect success in encounters, such as age, sex, and size. Nevertheless, it is clear in both cases that receivers respond to the signal in a way that we can assume is advantageous to the signaler. Similar, if sometimes less dramatic, effects have been shown by manipulating badges in dark-eyed juncos (Holberton et al. 1989, Grasso et al. 1996), great tits (Lemel and Wallin 1993), and house sparrows (Gonzalez et al. 2002). Some evidence also exists that enhancing badges affects success in aggressive competition for breeding resources, for example in collared flycatchers (Qvarnström 1997).

All the above cases, in which dominance was enhanced by enhancing a badge, involve badges based on melanin pigments. Initial work with carotenoid-based plumage ornaments, for example in northern cardinals (*Cardinalis cardinalis*) (Wolfenbarger 1999) and house finches (McGraw and Hill 2000), indicated that these signals do not affect dominance. Recently, however, carotenoid-based ornaments have been shown to function as badges of status in red-shouldered widowbirds (*Euplectes axillaris*). Males of this species have red epaulets that are produced by carotenoid pigments. Pryke and Andersson (2003) manipulated the size and the redness of epaulets in both captive and free-living males. In captivity, the redness and especially the size of the epaulet affected the likelihood of winning aggressive interactions at feeders; males with enlarged red epaulets, for example, won 93% of their contests against males with reduced red epaulets. In the field, males with enlarged red epaulets were significantly more likely than controls to acquire territories, whereas males with reduced red or reduced orange epaulets were significantly less likely to acquire territories. In a related species, the red-collared widowbird (*Euplectes ardens*), another carotenoid-based ornament had similar effects: males given red collars dominated both males with orange collars and males without collars in captivity, and males with enlarged red collars acquired larger territories than control males when free-living (Pryke et al. 2002). Thus carotenoid ornaments can directly affect the outcome of male-male aggressive contests, just as melanin signals do.

In many of the experiments on signaling via badges, the effects of badge manipulation have been tested using individuals that are unfamiliar with one another prior to the experiment. This precaution is taken under the reasonable assumption that badges are less likely to affect dominance relations if individuals already know each other's fighting abilities. In addition, many of the badge-enhancement experiments have focused on the initial series of encounters between individuals after they first encounter one another, presumably with the idea that the effects of the badge may disappear once the individuals gain direct experience with each other's abilities. If badges indeed only affect the outcome of initial encounters between unfamiliar individuals, this would limit their importance as signals. Some evidence suggests, however, that these limitations do not apply, at least not in all cases. In the experiments on badge manipulation in free-living widowbirds, for example, males with enhanced badges were able to maintain possession of their larger-than-average territories for periods of many weeks, while at the same time experiencing fewer intrusions than did other treatment groups (Pryke et al. 2002, Pryke and Andersson 2003). Since neighboring territory owners in birds are generally able to recognize each other as individuals (Falls 1982), these results suggest that the effects of badges can persist even among individuals familiar enough to recognize each other individually.

RELIABILITY OF BADGES

In assessing the reliability of badges, we are as usual faced with the question of just what it is that the signal might be reliable about. One oft-cited answer is dominance. Badge size (or color, or contrast) has been shown to correlate with dominance rank (or some other measure of success in encounters) in a number of species (Rohwer 1975, Balph et al. 1979, Møller 1987a, Senar et al. 1993, Pärt and Qvarnström 1997, Hein et al. 2003). We see two problems in interpreting badges as signaling dominance, however. First, dominance is a relationship between two individuals rather than an attribute of one, so a signaler cannot really be said to have an intrinsic dominance that it can signal. Second, and more important, because the badge is hypothesized to affect dominance directly, through its impact on receivers, it seems awkward, and almost circular, to interpret badges as simultaneously signaling and causing dominance.

Instead, we think it makes more sense to interpret badges as signaling fighting ability. With this interpretation, an ideal test of reliability would be to take a group of birds varying in badge size and unfamiliar with each other, manipulate their badges to some standard form (e.g., give them all the average badge size), and then assess their success in fighting each other. If badges signal fighting ability, then we predict that the original, pre-manipulation badge size would correlate with success in encounters after the manipulation. Lacking data of this kind, the best we can do is to test whether badge size correlates with attributes known to affect fighting ability, such as age, size, and sex. In birds, adults tend to dominate juveniles, larger individuals tend to dominate smaller, and males tend to dominate females (though there are exceptions to all these generalizations). If badges signal fighting ability, then badge size should increase with age and body size, and should be larger in males than in females.

We have already mentioned instances in which badge size differs between age and sex classes. The expected relationships between age, sex, and badge size are well illustrated by Harris's sparrows. Rohwer et al. (1981) scored badge size in winter Harris's sparrow by comparing birds to a set of photo-

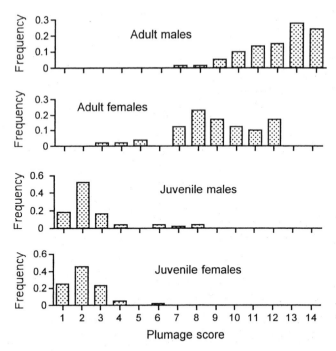

FIGURE 4.4. Variation in the amount of black plumage in the badges of Harris's sparrows of different age and sex classes, from Rohwer et al. (1981). Badges are scored for the amount of black, from 1 (least black) to 14 (most black). Adults receive higher scores than juveniles, and among the adults the males receive higher scores than females. Juvenile males and juvenile females do not differ significantly.

graphs, arranged from 1 (lowest amount of black) to 14 (most black). Adults of both sexes scored higher than did juvenile birds. Adult males had significantly larger badges than adult females, but there was no difference between males and females among the juveniles (figure 4.4). Badge size was also positively correlated with body size, measured as wing length, within adult males, though not within other age/sex classes.

In house sparrows, only males possess the badge, which is again a black throat patch. Within male house sparrows, Veiga (1993) found a significant correlation between badge size and age, with most of the increase occurring between ages one and two. Within first-year birds banded as nestlings, badge size was negatively correlated with fledging date; thus badge size increased with age in these juvenile males on the level of days and weeks. Gonzalez et al. (2001) found a significant, positive correlation between badge size and plasma testosterone level in house sparrows. This association is potentially important because testosterone often increases aggressiveness in birds, and increased aggressiveness in turn enhances fighting ability.

The size of a badge is not necessarily its most important attribute. In white-crowned sparrows, the signal appears to lie in the degree of contrast between the dark and light stripes of the crown. Fugle and Rothstein (1985) scored white-crowned sparrows on this contrast, using separate seven-step scales for first years and adults. These two age classes thus were assumed to be completely nonoverlapping in crown coloration. Within both age classes, males had significantly more contrasting crowns than did females. Correlations with size were not tested separately, though males are on average larger than females, and adults larger than first years.

Some authors have argued that, when badges turn out to vary mainly by age and sex, this result somehow undermines the status-signaling hypothesis (Whitfield 1987, Senar et al. 1993). We disagree; if age and sex are primary determinants of fighting ability, then age and sex are precisely what a status signal ought to convey. Correlations with age and sex would be destructive to status signaling only if one could maintain that birds are able to assess age and sex independently of badge size, and therefore do not attend to badge size when assessing fighting ability. The results of badge-manipulation experiments show that this conclusion is incorrect, at least in some cases. In fact, one might predict that badges will have evolved only in species in which age and sex are not easily assessed by other means, the badges thus providing information that is not otherwise readily available.

We conclude that badges do contain reliable information about fighting ability, which provides a sufficient explanation for why badge attributes affect receivers in aggressive contexts.

Costs of Badges

Attributes of badges, such as badge size, are reliably correlated with traits that contribute to fighting ability. But why are these correlations reliable? If large badges aid in winning aggressive encounters, why do not all individuals develop large badges? Many answers have been suggested since this question was first posed by Rohwer (1975). These hypotheses can be divided into two categories, those that propose that winning encounters is not really advantageous, that is, that subordinates are as equally fit as dominants, and those that propose that reliability is maintained by signal costs, that is, that large badges have some cost that bears more heavily on individuals of low fighting ability.

What distinguishes the first of these two categories of hypotheses is that a disadvantage is proposed for dominance itself rather than for large badge size. Dominants have at least one advantage: they win aggressive encounters (which is what defines them). Aggressive encounters occur over resources, and so dominants ought to garner more resources, at least in the short term, than subordinates do. If dominants are to have equal fitness with subordinates, some disadvantage must balance this advantage. One possibility is that because dominants require higher metabolic rates, they burn up resources at a faster pace than subordinates do. Metabolic rate does increase with dominance in some status-signaling species (Røskaft et al. 1986, Hogstad 1987), but the opposite appears true in at least one case (Senar et al. 2000). A second idea, termed by Rohwer and Ewald (1981) the "shepherds hypothesis," is that dominants have a disadvantage in having to contend with other dominants for the privilege of residing in good habitat. Subordinates, by contrast, are tolerated by dominants because of their value as finders of food, and therefore pay lower costs of aggression. Dominants play the shepherds and subordinates the sheep. As evidence for the hypothesis in Harris's sparrows, Rohwer and Ewald showed that birds similar in dominance rank (as judged by badge size) contested with each other more if dominant than if subordinate.

Although proponents of the equal-fitness hypotheses have suggested some plausible costs to being dominant, we are skeptical of the conclusion that any such costs fully balance the benefits of dominance. Dominants, after all, are phenotypically superior in some aspects: they tend to be older, to be larger, and to have greater fighting ability. It seems logical that they would reap a net fitness benefit from this superiority. (This argument does not apply in comparing males and females, which typically have equal fitness on average within species [Fisher 1930].) Moreover, studies of winter birds that have attempted to integrate over fitness costs and benefits by measuring survival indicate that dominants usually have higher fitness than subordinates (Ekman and Askenmo 1984, Lahti 1998).

Hypotheses in the second category propose costs for the signal—the badge—rather than for dominance. One obvious possibility is that large badges have a receiver-dependent cost, because they elicit attacks by other individuals, leading to what Rowher (1977) termed "social control" of the signal. Hurd's (1997) model of aggressive signals incorporates such a receiver-dependent cost. The model assumes that a badge indicating strength is costly because it provokes attacks by strong individuals. This cost is especially high for weaker individuals because it is assumed that being attacked by an individual of greater strength is more costly than being attacked by one of equal strength. The model thus predicts that weak individuals with large badges (i.e., cheaters) are attacked more often by dominants than are honest subordinates, and that attacks by dominants are more costly to subordinates than are attacks by other subordinates. The first prediction is sometimes generalized to say that "like versus like" aggression should be particularly common, that is, aggression between individuals with similar badges.

The best way to test predictions of the social-control hypothesis is to create cheaters by artificially enhancing the badges of selected individuals and then put these experimental cheaters together with a mix of individuals having unaltered badges. The largest-scale such experiment was done by Fugle and Rothstein (1987), using white-crowned sparrows. Remember that in this species the status signal lies in the degree of contrast of the dark and light crown stripes. Fugle and Rothstein (1987) chose to manipulate the badges of first-year females, the group at the bottom of the status hierarchy. Cheaters were created by painting the crowns of some immature females with the bold black and white stripes of adult males. Controls were immature females that either were untreated or had their crowns painted to resemble their original coloration. In one set of trials, 20 cheaters and 30 controls were observed in free-living flocks after their release back into the wild at two sites. Neither the cheaters nor the controls were able to beat adults after treatment, but the cheaters did have a significantly better winning record than the controls did against other immatures. Contrary to our first prediction, adults attacked cheaters no more frequently than they attacked controls. Controls were attacked by other immatures more frequently than were cheaters. The second prediction was not tested quantitatively, but Fugle and Rothstein noted no obvious increase in the intensity of aggressive interactions involving the cheaters. In parallel experiments with captive flocks, adults if anything attacked cheaters less often than they attacked controls. Predictions of the social-control hypothesis thus fail miserably for this species.

Another negative case is provided by house sparrows. Gonzalez et al. (2002) experimentally increased badge size in low-ranking birds and then returned them to their captive flocks. The experimental birds were not involved in more fights overall after manipulation than before, nor were they involved in proportionately more fights with the highest-ranking birds. Furthermore, plasma corticosterone levels, a commonly used measure of stress, did not go up after the increase in badge size, indicating that the cheaters were not experiencing severe enough aggression to raise their levels of stress. The social-control predictions thus fail again.

Rohwer (1977) obtained very different results with Harris's sparrows. Rohwer captured birds with small badges from a winter flock, enlarged their badges using either magic marker (N = 8) or hair dye (N = 1), and then released them back into the flock. Uncaptured, unmanipulated subordinates served as an informal control. Relative to the controls, the experimentals suffered a significantly higher proportion of attacks from dominants after treatment than before. (The intensity of attacks was not reported.) Thus our first prediction was confirmed: cheaters suffered a cost due to being attacked more often by dominants. Møller (1987b) reported similar results for captive flocks of house sparrows; note that this outcome conflicts with Gonzalez et al.'s (2002) results with the same species.

For Harris's sparrows there is some evidence that aggression is particularly common among individuals similar in badge size (Rohwer and Ewald 1981). In contrast, white-crowned sparrows are more likely to attack those below them in rank than to attack equals (Keys and Rothstein 1991). This difference in the relative frequency of like-versus-like aggression may provide a proximate explanation for why badges appear to have receiver-dependent costs in Harris's sparrows and not in white-crowned sparrows. Receiver-dependent costs are more likely to be incurred if like-versus-like aggression is more common than like-versus-unlike aggression, although this begs the question of what other factors contribute to the different levels of like-versus-like aggression observed in these two species. Like-versus-like aggression also occurs more often than expected in another status-signaling species, the dark-eyed junco (Balph et al. 1979), so receiver-dependent costs may be expected in this species as well.

Veiga and Puerta (1996) suggested that badges may have significant production costs, contradicting the usual assumption that melanin-based plumage traits are cheap to produce. As a test, they gave juvenile house sparrows unlimited food before the autumn molt and found that the birds produced larger badges than are seen in free-living juveniles. This result provides only weak evidence for an effect of nutrition, however, since many factors other than diet must have differed between the experimental birds and the free-living comparison group.

More carefully controlled experiments have produced negative results on the importance of nutrition. Gonzalez et al. (1999) manipulated protein content of the diet of captive juvenile house sparrows over a period that spanned the autumn molt, and found no effect on the size or color of the badges that the birds produced. McGraw et al. (2002) nutritionally stressed a group of juvenile male house sparrows by preventing access to food during unpredictable periods, again over a period that spanned the autumn molt. The stressed group produced badges that were no different in size or color than the badges of unstressed controls. Overall, then, the assumption that melanin-based badges lack important production costs seems justified.

Another proposed production-related cost of badges involves the link between testosterone's role in badge expression and its effect on immune function. Recent evidence shows that testosterone enhances the size of badges in house sparrows. Natural variation in plasma testosterone levels prior to a molt correlates positively with the increase in badge size during the molt (Evans et al. 2000, Gonzalez et al. 2001). In addition, males treated with large doses of testosterone prior to molting exhibit larger increases in badge size during the molt than do untreated males (Evans et al. 2000, Buchanan et al. 2001). This testosterone dependence of badge size raises the possibility that the cost of large badges is an immunological one, for testosterone has been proposed to depress immune function (Folstad and Karter 1992). The badges of house sparrows would then be an example of an "immunocompetence handicap"—a signal of quality that is reliable because only individuals of high quality can afford to pay the costs of lowered immune defense.

Whether testosterone has an immunosuppressive effect is controversial, however. Some experiments with birds have found the predicted negative effect of testosterone treatment on aspects of immune function or parasite resistance (Duffy et al. 2000, Peters 2000, Saino et al. 1995, Duckworth et al. 2001), whereas others have not (Ros et al. 1997, Hasselquist et al. 1999, Lindström et al. 2001). For our purposes, studies of immunocompetence in house sparrows are most relevant, since this is the species in which testosterone has been shown to enhance badge size. Poiani et al. (2000) measured ectoparasite abundance on male house sparrows that had been castrated and treated with varying levels of testosterone. Males given large implants, sufficient to produce testosterone levels at the high end for breeding birds, showed a substantial increase in ectoparasite loads relative to untreated controls. In other studies, high levels of testosterone produced some suppression of antibody production in response to a novel antigen (Evans et al. 2000, Buchanan et al. 2003a), but had no effect on cell-mediated responses (Buchanan et al. 2003a).

Although there is some evidence of an immunosuppressive effect of testosterone in house sparrows, the positive results all involve testosterone levels typical of the breeding season. Badge size, however, is influenced not by testosterone during the breeding season but by testosterone at the time of molt (Buchanan et al. 2001), when levels are typically on the order of 20 times lower than during breeding (Hegner and Wingfield 1986). Manipulation of testosterone within the range found during the molt produced no evidence of a suppressive effect on either antigen production or cell-mediated responses (Buchanan et al. 2003a). It thus seems likely that house sparrows, at least, can produce a large badge without paying any cost associated with immunosuppression.

Predation has been suggested as a final possible cost of badges. Large or conspicuous badges may attract the attention of predators, and if weak individuals are less able to survive the increased attention than are strong ones, then predation could serve as a cost that stabilizes the signaling system. Data on badge size and predation are few, however. Veiga (1995) found some evidence of increased mortality among first-year house sparrows with enhanced badges, which is consistent with this cost.

CONCLUSIONS

Status signaling has been studied most extensively in white-crowned sparrows, Harris's sparrows, and house sparrows. In all three species, individuals with experimentally enhanced badges experience an advantage in aggressive encounters, at least over others of the same age and sex. In all three species, we can understand why receivers attend to the badge, for the characteristics of the badge reliably correlate with traits of the signaler that are important in settling contests, such as age, sex, and size. As Rohwer (1975) pointed out long ago, the most difficult aspect of these systems to understand is what it is that prevents cheating, that is, the development of a larger badge than one deserves. For Harris's sparrows, an answer appears to be available: larger badges elicit

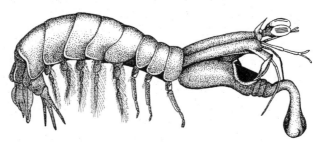

Figure 4.5. A stomatopod of the genus *Gonodactylus*. Note the black meral spot on the enlarged fourth segment (the merus) of the raptorial appendage.

increased aggression from birds of superior fighting ability. This gives a cost of the sort that is capable of stabilizing such a signaling system—a cost that falls more heavily on individuals of lower quality. For white-crowned sparrows and house sparrows, however, the evidence is against such a receiver-dependent cost for the most effective badges, and yet the signaling system still functions. Since the signaling system appears to be just as stable without receiver-dependent costs as with them, this particular type of cost does not seem to be necessary to the maintenance of a signaling system.

Other kinds of costs may instead apply, ones that are not receiver-dependent. The possibility of an immunosuppressive cost is particularly intriguing, but there are problems with this idea. The hypothesis assumes that testosterone is important to the production of badges, which has been shown only in one species, and that testosterone is immunosuppressive, for which the evidence is mixed. Further work may well shore up these aspects of the immunosuppression hypothesis, but it will still be questionable whether production of a badge requires that testosterone levels be elevated long enough and high enough to have important immunosuppressive effects.

Cheaters have been experimentally produced in status-signaling systems, but we do not know whether they occur naturally. Thus, the extent to which deception occurs in status signaling is unknown. One way to begin looking for deception would be through the experiment we suggested for measuring badge reliability: obtain a sample of birds varying in original badge size and unfamiliar with each other, standardize their badge sizes, determine their fighting abilities, and correlate fighting ability with original badge size.

Weapon Displays in Crustaceans

Many species of crustaceans have enlarged appendages used in fighting. In stomatopods, for example, the second pair of maxillipeds has been lengthened and strengthened to produce powerful weapons, the "raptorial appendages," which are used in both prey capture and fighting (figure 4.5). A subset of the stomatopods, species that Caldwell and Dingle (1975) have termed "smashers," use their raptorial appendages to disable armored prey such as mollusks and crabs. When these weapons are used against conspecifics, they are capable of causing serious injury, even death. In another group, the snapping shrimps, the claw (or chela) of one of the first pair of walking legs is greatly enlarged. This claw can be closed rapidly to produce an audible snap. A snap produced against the body of a conspecific is capable of causing severe injury (Knowlton and Keller 1982).

Not surprisingly, crustaceans often use these and similar weapons in aggressive displays; the weapons are extended, waved, or otherwise brandished during agonistic encounters. Such displays may function to signal aggressive intentions; in particular, brandishing a weapon may signal that attack is imminent. A weapon display also may serve to signal fighting ability, if weapon size is important in defeating opponents, or if weapon size correlates with body size and body size is important in defeating opponents.

The idea that weapons function to signal fighting ability has also been suggested for other categories of weapons, such as the antlers of deer (Clutton-Brock 1982) and the horns of sheep (Geist 1971). Clutton-Brock (1982), however, has argued that the empirical evidence for a signal function of weapons is slim in all these groups. We believe that the evidence is stronger for crustaceans, which is one reason for concentrating on these animals; the other is that weapon displays in crustaceans provide some of the best evidence of deception available for any type of aggressive signal.

Response to Weapon Displays in Crustaceans

Stomatopods employ a weapon display termed the "meral spread." Here, the largest segments of the raptorial appendages, the meri, are spread outward, revealing the size of the appendages and exposing conspicuous "meral spots" on their inner surfaces (Dingle and Caldwell 1969). The meral spread is used in aggressive encounters, which are frequent in stomatopods, occurring especially during competition for dwelling cavities. Dingle (1969) staged encounters between same-sex pairs of captive stomatopods of the species *Gonodactylus bredini*, and used his observations to test whether the production of a meral spread by one animal affected the immediately following behavior of the other. During the first 10 minutes of encounters, opponents were significantly more likely to avoid the signaler, and significantly less likely to strike or chase, after a meral spread than after some other behavior. Reaction to meral spreads was similar later in contests. Meral spreads thus appear to be effective in eliciting avoidance and deterring attack by opponents, though the evidence here is only correlational.

Snapping shrimp use a weapon display, termed the "open-chela display," in which the enlarged, major chela is cocked open as if about to be snapped.

Evidence suggests that this display is used to signal size, which is highly important in determining the outcome of aggressive interactions (Hughes 1996). Hughes (1996) presented major chelae, separate from the rest of the body, to captive snapping shrimp of the species *Alpheus heterochaelis*, using chelae of various sizes taken from molted exoskeletons, and noting the number of open-chela displays given by the test subjects in response. For male snapping shrimp presented with an open chela, the level of response depended on the size of their own chelae relative to the test chela: males gave more displays as this ratio (own/test) increased. No such relationship occurred when the test chela was presented closed. One interpretation of these results is that males assess an opponent's fighting ability on the basis of the size of his open chela, compare this assessment to their own ability, and react more aggressively (i.e., with more aggressive displays) the higher they assess their own ability relative to the other's. Whether or not this sophisticated level of assessment occurs, it is clear nonetheless that the size of the weapon is used as a signal, for the animals react to weapon size when the weapon is presented divorced from any other cue. What remains to be demonstrated is that a large signal is effective in the sense of contributing to victory in encounters.

Reliability of Weapon Displays in Crustaceans

The meral-spread display of stomatopods has mainly been interpreted as a signal of intention rather than a signal of fighting ability, whereas the reverse has been true of the open-chela display of snapping shrimp. It seems likely that in reality both displays signal both categories of information, but existing evidence bears on the simpler interpretations.

Evidence for the reliability of the meral spread as a signal of intention comes from laboratory encounters staged by Dingle (1969). Dingle examined within-individual sequences of behavior during aggressive encounters, and assessed whether one act, such as a meral spread, was followed more or less often than expected by other specific acts. Unfortunately for our purposes, Dingle eliminated from consideration nonaggressive behaviors, such as "avoids" and "does nothing." Because the list of following acts was restricted to aggressive behaviors, what the analysis tells us is whether the meral spread predicts one kind of aggressive act rather than another, and not whether the display predicts aggression rather than nonaggression. Dingle found that meral spreads were followed more often than expected by lunges, less often than expected by meral spreads, and about as often as expected by strikes and chases. If we assume that a lunge is more dangerous to an opponent than is a meral spread, and less dangerous than is a strike or chase, then we can take these results to mean that meral spreads signal a moderate escalation of aggression. This interpretation might have been otherwise, however, if nonaggressive acts had been included in the analysis.

Hughes (1996, 2000) interpreted the open-chela display of snapping shrimp as communicating fighting ability. In her study species, *Alpheus heterochaelis*, the larger of two animals (measured in terms of body length) won 88% of aggressive encounters staged between same-sex individuals, and the probability of victory increased as the larger animal's size advantage increased (Hughes 1996). Of 13 encounters in which one individual had the larger body size and the other had the larger chela, nine were won by the animal with the larger body compared to only four by the animal with the larger chela (Hughes 2000). Body size, rather than chela size, thus seems to be the primary determinant of fighting ability. Chela size was strongly correlated with body size in both males ($r^2 = 0.82$) and females ($r^2 = 0.72$) (Hughes 1996). Thus chela size, when revealed in an open-chela display, signals body size and hence fighting ability with considerable, but not perfect, reliability.

COSTS AND DECEPTIVE USE OF WEAPON DISPLAYS

Raptorial appendages and major chelae both must be quite costly to produce, simply judging by the size of these structures relative to the rest of the body. But the structures have additional functions, in fighting and (in the case of raptorial appendages) in prey capture, and their production costs thus cannot be totally, or even largely, ascribed to signal costs. Whether there are significant energetic costs associated with the displays in which these weapons are brandished has not been studied. The most likely costs for weapon displays in general would seem to be receiver-dependent costs. In stomatopods, the meral-spread display actually lowers, rather than increases, the chances that an opponent will attack the signaler, at least when the signaler has not recently molted. What happens when the signaler has recently molted leads us to the intriguing evidence for deception in this group.

During the period just after a molt, when the new cuticle is still hardening, stomatopods are particularly vulnerable to injury because the exoskeleton no longer serves as armor. Concomitantly, the softness of the cuticle renders the raptorial appendages ineffective as weapons; in fact a newly molted animal delivering a blow injures itself rather than its opponent (Adams and Caldwell 1990). Some newly molted stomatopods nevertheless continue to give threat displays as if they were still dangerous; it is these "bluffing" displays in newly molted animals that have been suggested to be deceptive.

Steger and Caldwell (1983) allowed newly molted individuals of the stomatopod *Gonodactylus bredini* to establish residency in cavities in the lab. Individuals between molts ("intermolts") were then introduced to contest for cavity ownership. Most newly molted residents either hid (N = 30) or fled (N = 13), but some actively defended their cavities by displaying (N = 17). Active defense was considerably more common among control, intermolt residents (19/25 actively defending). Of the newly molted residents choosing active defense, two lunged at the opponent and 15 gave a meral spread. By contrast, only four of 19 control residents gave meral spreads, most choosing instead to strike. Newly molted individuals thus adjust their contest behavior to their circumstances, emphasizing display and de-emphasizing direct attacks. Further adjustment occurs when new-molts are confronted with intermolt opponents of different sizes. Adams and Caldwell (1990) found that the number of threat displays given by newly molted residents increased as their size increased relative to opponent's size. The opposite was true for intermolt residents, which gave fewer threats as their size increased relative to opponent's size, presumably because large intermolt residents were more likely to proceed directly to attack.

The meral-spread display should probably be interpreted as simultaneously signaling fighting ability and intention to escalate. A meral spread given by a newly molted individual, then, is deceptive, in the sense that the signaler has both lower fighting ability and less intention of escalating than implied by the signal. Adams and Caldwell (1990) found that the deceptive displays of newly molted individuals affected the behavior of intruders. Intruders were significantly less likely to escalate their own aggressive behavior if a newly molted resident threatened than if it did not. Furthermore, bluffing was effective, in the sense that bluffers were more likely to retain their cavities than were newly molted individuals that remained with the cavity but did not display (figure 4.6; Adams and Caldwell 1990).

The evidence for a cost of bluffing in this system is ambiguous. Adams and Caldwell (1990) considered a newly molted resident to be at high risk if an intruder entered the cavity while the resident was still inside. In nature, the resident is usually killed in this situation, though Adams and Caldwell (1990) prevented that outcome in their experiments. Newly molted residents that remained with the cavity and threatened were more likely to end up at high risk than were newly molted residents that fled, but less likely than were newly molted individuals that remained with the cavity and did not threaten. Bluffing thus is costly relative to fleeing, but not relative to remaining with the cavity and failing to threaten. Why the latter tactic is retained is not clear, since it appears to be both costly and ineffective. A partial explanation is that some of the newly molted residents fail to detect an intruder until it has advanced far enough to block the cavity entrance, making it impossible for the resident to give a meral spread (Adams and Caldwell 1990). Nevertheless, some newly molted residents showed clear evidence of having seen the intruder, by tracking it with their eyestalks, and yet did not threaten (Adams and Caldwell 1990).

Intermolts may be able to exaggerate their aggressiveness or fighting ability to some degree, but outright deception can be performed only by newly molted animals. As a consequence, the frequency of outright deception is inherently limited by the life history and demographics of the species, and some degree of signal reliability is guaranteed. From the relative abundances of intermolt and newly molted individuals and the relative frequency at which each type

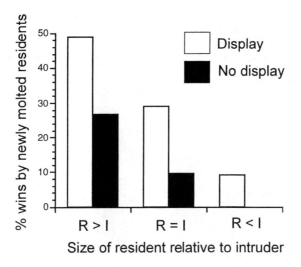

FIGURE 4.6. The percentage of newly molted stomatopods able to retain their cavities depending on whether they did or did not give meral-spread displays, from Adams and Caldwell (1990). Production of a meral-spread display by a newly molted stomatopod can be considered a bluff, in the sense that the animal appears to be threatening to strike when it is in fact unable to do so. Success in defending the cavity depended on the size of the resident relative to that of the intruder, but for each category of size relationship the residents were more successful when they bluffed than when they did not. The advantage of display was statistically significant for cases where the resident was larger than the intruder (R > I) or equal in size (R = I) but not for cases in which the resident was smaller (R < I).

gives meral spreads, Caldwell (1986) estimated that 85% of residents giving a meral spread are intermolts. If receivers are unable to discriminate newly molted animals from intermolts, then the existence of this large pool of honest signalers explains why receivers continue to attend to the signal—they attend because much of the time the resident giving a meral spread actually is dangerous. Caldwell (1986) suggested that resident stomatopods pair meral spreads and strikes during the intermolt period in order to reinforce the message of the display and build a reputation for aggressiveness that will carry over into the period after molt. Stomatopods are able to recognize individuals by odor, and they avoid cavities of residents that have previously defeated them (Caldwell 1984). As predicted, aggressiveness increases in residents immediately prior to molt, and they are particularly likely to pair meral spreads and strikes during this period (figure 4.7; see also Caldwell 1986, table 7-3).

One reason that the case for deception in stomatopods is so convincing is that a clear qualitative difference exists between dishonest and honest signalers: dishonest signalers have a soft cuticle and honest signalers a hard one. Because of this discrete difference, we can identify deceptive individuals with-

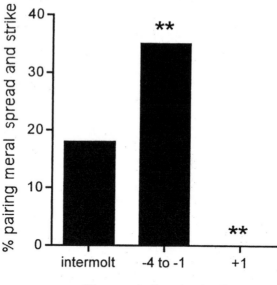

FIGURE 4.7. Pairing of meral spreads with strikes relative to the onset of molt in the sto-matopod *Gonodactylus bredini*, from Caldwell (1986). Individuals are more likely to pair a meral spread with a strike in the period leading up to the molt (days −4 to − 1) than during the intermolt period; this pairing may condition others to fear the meral-spread display when it is given after the molt by the same individual. Immediately after a molt (day +1) no strikes are delivered. ** indicates a significant difference from the in-termolt period (P < 0.01) by a chi square test.

out ambiguity. A parallel case with two discrete classes of signalers occurs in the fiddler crab *Uca annulipes* (Backwell et al. 2000). Here, males that lose their major claw regenerate a new one that is only 80% the mass of an original claw of the same length. The lighter, regenerated claws seem to be inferior to original claws, for males with regenerated claws are considerably more likely to lose forced encounters in the lab (Backwell et al. 2000). Whether opponents can be bluffed by regenerated claws has not yet been tested, but this seems to be another promising system to examine for deception.

A more subtle type of deception may also be occurring in some crustacean groups, deception in which signalers that vary quantitatively in fighting ability, rather than qualitatively, signal a slightly exaggerated position along the con-tinuum of ability. Hughes (2000) suggested that this kind of subtle deception occurs in the snapping shrimp *Alpheus heterochaelis*. As we have seen, body size is extremely important to success in fights in this species, the size of the major chela is strongly correlated with body size, and receivers respond to the size of the major chela as revealed in the open-chela display. The correlation

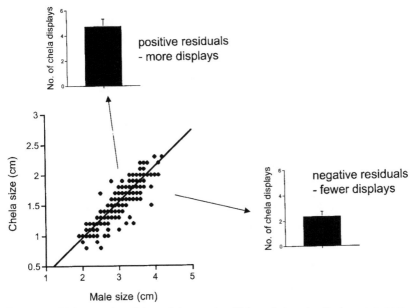

FIGURE 4.8. Male snapping shrimp of the species *Alpheus heterochaelis* give more open-chela displays when their chelae are larger than expected for their body sizes, from Hughes (1996, 2000). The scattergram shows male chela size versus body size, together with the regression line. Males with positive residuals (and thus larger chelae than expected for their body sizes) gave significantly more displays in 3 minutes than did males with negative residuals (and thus smaller than expected chelae).

between body size and chela size naturally is not perfect. If chela size is regressed on body size, some individuals have a chela size somewhat above that predicted by their body size (a positive residual), whereas others have a chela size somewhat below that predicted (a negative residual) (figure 4.8). Individuals with positive residuals in a sense have an exaggerated signal of their size and thus of their fighting ability. As evidence that these animals exploit this exaggeration, Hughes (2000) showed that animals with positive residuals give open-chela displays more frequently than do animals with negative residuals (figure 4.9). Interactions involving animals with positive residuals tend to last longer, suggesting that the signal size has some effect on opponents. What has not been shown in this system is whether an exaggerated chela size is effective in winning contests.

CONCLUSIONS

The production of meral-spread displays by newly molted stomatopods is often cited as the clearest example of intraspecific deception currently

FIGURE 4.9. A calling male cricket frog.

known. Again, what makes this example so clear is the discrete difference in aggressiveness and fighting ability that exists between newly molted and intermolt individuals. If we accept that the meral spread is a signal of aggressiveness, fighting ability, or both, then we can conclude that newly molted individuals are signaling false information. If we accept the correlational evidence that newly molted individuals are more likely to retain their cavities as a result of giving the meral spread, then we can conclude further that newly molted individuals get a selective benefit from giving the display. Meral spreads by newly molted stomatopods then meet both our criteria for deception.

The occurrence of deceit in the meral-spread display is well explained by game-theory signaling models, in particular by the model of Adams and Mesterton-Gibbons (1995), which was formulated with this system in mind. The model assumes that signal costs are receiver-dependent, that weak individuals are more likely to pay these costs because they are more likely to lose, and that, when paid, the costs are greater for weak individuals than for strong ones. All of these assumptions accord with the meral-spread system, in which newly molted individuals are both likely to lose and highly vulnerable to injury. In the Adams and Mesterson-Gibbons model, the aggressive display is given by both the strongest and the weakest individuals; the latter display, despite paying heavy costs, because they receive especially large benefits from displaying. In stomatopods, the weakest individuals, that is, the newly molted ones, do seem to benefit disproportionately, because displaying gives them their only opportunity to retain their cavity, which is extremely important to their fitness.

Of course, the model did not predict the properties of this signaling system in advance, but rather was formulated to explain them post hoc; nevertheless, the model provides a convincing explanation for the evolutionary stability of aggressive signaling in stomatopods.

The case for deception in the open-chela display of snapping shrimp is very different, in the sense that the distinctions between honest and deceptive signalers are small and continuous, rather than large and discrete. In snapping shrimp, deception consists of having a larger chela for one's body size than predicted by the general chela size versus body size relationship. Animals with larger than predicted chelae definitely exist, and can be taken as signaling false information. A weakness in this case is that it has not been shown that these animals benefit from signaling falsely. Nevertheless, the evidence that those individuals with larger than expected chelae signal more than others provides support for the hypothesis that they are benefiting disproportionately, and that this higher level of signaling represents deception by our definition.

Dominant Frequency in Calls of Frogs and Toads

During the breeding season, males of many species of frogs and toads produce loud, obvious vocalizations termed "advertisement calls" (Wells 1977). These vocalizations typically serve dual functions in attracting females for mating and in warding off and intimidating rival males (Gerhardt 1994). Advertisement calls thus are analogous to the songs of acoustic insects and passerine birds (Searcy and Andersson 1986, Bailey 1991). What is particularly interesting about anuran advertisement calls is that a single property of the calls, their dominant frequency, conveys information about a trait of overwhelming importance in resolving aggressive contests, namely body size.

"Dominant frequency" in these studies is defined as the acoustic frequency with the greatest energy in the signal. Within many species of frogs and toads, the dominant frequency of the advertisement call is inversely correlated with the body size of the caller; in other words, the largest males give the deepest croaks. Another widely recognized fact is that larger body size is of great advantage in winning fights in many anuran species (e.g., Davies and Halliday 1978, Howard 1978, Arak 1983, Robertson 1986). The dominant frequency of an anuran call depends to some extent on the size, especially the weight, of the vocal cords (Martin 1971, 1972), and the size of the vocal cords must in turn be constrained by overall body size. The argument thus runs that dominant frequency depends on size of vocal cords, size of vocal cords depends on body size, and body size determines fighting ability; therefore, dominant frequency is constrained to be an honest signal of fighting ability. In the following, we consider the extent to which this constraint actually operates.

RESPONSE TO DOMINANT FREQUENCY

Davies and Halliday (1978) provide an experimental demonstration, for the toad *Bufo bufo,* that the dominant frequency of a male's call affects the behavior of opponents in male-male encounters. In the lab, a small male was allowed to grasp a receptive female in amplexus (the anuran mating posture), and a second male was introduced into the tank to try to supplant the first. When such encounters occur in nature, the defending male always calls when attacked, but in this experiment the defending male was prevented from calling by means of a rubber band stretched behind the arms and through the mouth ("like a horse's bit"). In place of the defender's own calls, the researchers played calls recorded from either a small or a large male, using a loudspeaker placed over the tank. Calls were played whenever the supplanting male contacted the mating pair. Attacks by the supplanter were over three times more frequent when calls of small males were played than when calls of large males were played (Davies and Halliday 1978). The effect of call type was much greater in the trials with small defenders than in trials with large defenders; in the latter, attacks were rare regardless of which call was played, so the second male must have some way of assessing the size of the defender independent of the call. Nonetheless, it is clear that in trials with small defenders, calls of large males were effective in deterring opponents.

In this experiment, the calls of large males had a lower dominant frequency than the calls of small males, but we cannot be sure that this was the sole call parameter that differed with caller size. Therefore, we cannot be certain that dominant frequency was the parameter to which attacking males attended. Arak (1983), working with a second toad species, the natterjack toad (*Bufo calamita*), eliminated any uncertainty about which call parameters were important to receivers by using for playback artificial calls that were identical in all respects except for their dominant frequency. Two calls were synthesized, using dominant frequencies characteristic of the largest and smallest males in the study population, respectively. These calls were then broadcast to calling individuals in the field. The percentage of males swimming away from the loudspeaker was significantly greater for playback of the large-male call (51%) than for playback of the small-male call (8%); conversely, the percentage of males attacking the loudspeaker was significantly greater for the small-male call (61%) than for the large-male (20%). Wagner (1989) obtained similar results in Blanchard's cricket frog (*Acris crepitans blanchardi*) (figure 4.9), again using synthetic calls: calling males in the field tended to abandon calling and retreat from speakers playing calls of low frequency, and to continue calling and attack speakers playing calls of high frequency.

The results of Davies and Halliday (1978), Arak (1983), and Wagner (1989) are particularly clear in showing not only that male frogs and toads discrimi-

TABLE 4.1.
Correlations between male size and call frequency in various species of frogs and toads.

Species	r	r²	N	P	Size Measure*	Reference
Acris crepitans	−0.47	0.22	238	0.001	SVL	Wagner (1989)
Bufo americanus	−0.30	0.09	26	NS	SVL	Zweifel (1968)
Bufo americanus	−0.49	0.24	99	0.001	SVL	Howard and Young (1998)
Bufo bufo	−0.88	0.78	20	0.001	SVL	Davies and Halliday (1978)
Bufo calamita	−0.57	0.33	21	0.01	SVL	Arak (1983)
Bufo fowleri	−0.59	0.35	41	0.001	SVL	Zweifel (1968)
Hyla cinerea	−0.80	0.64	81	0.01	SVL	Oldham and Gerhardt (1975)
Hyperolius marmoratus	−0.63	0.40	26	0.001	SVL	Passmore and Telford (1983)
Physalaemus pustulosus	−0.53	0.28	136	0.01	SVL	Ryan (1980)
Rana clamitans	−0.84	0.70	83	0.01	SVL	Bee et al. (1999)
Rana virgatipes	−0.87	0.75	69	0.0001	Mass	Given (1987)
Uperoleia rugosa	−0.71	0.50	100	0.001	Mass	Robertson (1986)

* SVL is snout-vent length.

nate between the calls of large and small males, but also that they respond to the large-male calls in a way that ought to benefit the caller (see also Robertson 1986). Other studies have shown a similar discrimination, but have used a response that is more ambiguous in terms of benefit to caller. Bee et al. (1999) found that male green frogs (*Rana clamitans*) increased their own calling rate more in response to synthetic calls of low frequency than in response to calls of high frequency. Similarly, male carpenter frogs (*Rana virgatipes*) gave aggressive calls at a higher rate in response to playback of low-frequency calls than in response to high-frequency calls (Given 1999). Clearly, receivers responded to the frequency of the signal in these last two species, but we cannot assume that a signaler would benefit from eliciting the observed responses.

Receiver response to dominant frequency in male frogs is not universal. Territorial male bullfrogs (*Rana catesbeiana*), for example, show equal response to synthetic calls of low and high dominant frequencies (Bee 2002), even though they are well able to perceive the frequency differences distinguishing these calls (Bee and Gerhardt 2001).

RELIABILITY OF DOMINANT FREQUENCY

Correlations between body size and dominant frequency of advertisement or aggressive calls have been measured in a substantial number of species of frogs and toads. A sample of these correlations is provided in table 4.1. The correlations are all negative, meaning that dominant frequency always goes down as body size goes up. Published correlations are almost entirely statistically significant, though there may be some bias toward not publishing insignificant correlations. Clearly, the preponderance of evidence demonstrates that information about the size of the caller can be extracted from call frequency. The median correlation coefficient, however, is only about −0.6, which corresponds to a coefficient of determination (r^2) of about 0.36. On average, then, only about 36% of the variation in body size can be explained by call frequency (and vice versa). Admittedly, some of the correlations are tighter, but even in these cases a male can have a body size substantially below or above that predicted by the frequency of his call. This raises the question of whether the remaining variation is simply error, caused by some combination of faulty measurement and environmental fluctuations beyond the control of the caller, or whether males can take advantage of some of this variation to present themselves as having a larger or smaller size than is actually the case.

To answer this question, we must first consider the nature of the vocal-production mechanisms that could influence dominant frequency. Martin (1971, 1972) investigated these mechanisms, in toads of the genus *Bufo*, by substituting artificial airflow for the pulmonary airflow that normally causes the vocal cords in the larynx to vibrate and thus generate sound. One might expect the length of the vocal cord to be of primary importance in determining fundamental frequency, but in fact Martin (1971) found no relationship between vocal-cord length and vocal-cord resonant frequency in a sample of 79 individuals of 33 taxa. Instead, he found a tight relationship between mass of the vocal cord and the fundamental frequency of resonance in this same sample. *Bufo* species have fibrous masses located near the center of their vocal cords, and Martin (1971) suggests that it is the presence of these masses that makes the relationship between mass and frequency so tight in this group. Presumably, the size of these masses can vary independently of body size to some extent; partial independence requires only that the correlation between body size and size of fibrous mass be less than perfect. The size of the fibrous mass at the center of anuran vocal cords would then explain some of the residual variation in dominant frequency that is not explained by body size.

Martin (1971, 1972) identified two additional factors that affect dominant frequency: pulmonary air pressure and vocal-cord tension. Martin (1971) found, for example, that the vocal cords of the Great Plains toad (*Bufo cognatus*) produce frequencies of about 1.6 kHz when driven at 20 mm HG of pressure and 2.3 kHz at a pressure of 180 mm HG. Lowering the dominant frequencies by lowering the driving pressure has a cost, however, because amplitude is also lower at lower pressures (Martin 1971). Vocal-cord tension is modified by the movement of cartilaginous valves in the larynx. The cartilage of these valves pushes against the vocal cords when the valves open, increasing tension and raising the resonant frequency of the cords (Martin 1971). The valves can be opened either by pulmonary air pressure or by muscle contraction (Martin 1971).

We are now in a better position to evaluate the idea that dominant frequency is constrained to be an honest indicator of body size. Because dominant frequency varies inversely with the mass of the vocal cords, and because the mass of the vocal cords can be expected to vary directly with body size (Martin 1972), we expect an inverse correlation between dominant frequency and body size, other things being equal. Other things are not necessarily equal, however, and mechanisms exist by which anurans can cheat on this relationship. On an evolutionary time scale, males can evolve more massive vocal cords for their body sizes, thus producing lower dominant frequencies at given body sizes. A particularly effective way of changing vocal-cord mass would be to increase the size of the fibrous masses at the center of the cords. Receiver perception might track changes in vocal-cord mass, so that at any one time males with average vocal-cord size for their body size would be perceived as having that body size. Still, males with above-average vocal-cord masses for their size (i.e., positive residuals) could benefit from producing lower dominant frequencies in a way that can be considered deceptive, similar to the kind of deception proposed by Hughes (2000) for snapping shrimp having larger chelae than expected for their body sizes. On a behavioral time scale, males could vary dominant frequency by varying pulmonary air pressure or vocal-cord tension; these mechanisms might certainly be considered deceptive. In summary, then, the mechanisms of call production in frogs and toads make it likely that dominant frequency will contain reliable information on caller size, but some level of dishonesty is by no means precluded.

Costs and Deceptive Use of Dominant Frequency

One way to signal exaggerated body size via dominant frequency would be to increase the mass of the vocal cords, particularly by increasing the size of the fibrous masses. No one, to our knowledge, has studied the fitness effects of such a strategy, but presumably this kind of exaggeration would produce some fitness benefits in male-male encounters. At the same time, we can assume that producing larger vocal cords or fibrous masses would have some developmental costs, but whether such costs would be more than negligible is unknown.

Alteration of pulmonary air pressure or vocal-cord tension should allow short-term, behavioral adjustment of dominant frequency, and evidence exists that these kinds of behavioral adjustments do occur in some species. Wagner (1989), for example, has shown that Blanchard's cricket frogs lower the dominant frequency of their own calls in response to call playback. Shifts in frequency are of greater magnitude for playback of low-frequency calls than for playback of high-frequency calls, and for the low-frequency calls shifts are greater for high-amplitude playback than for low. Males thus respond with greater shifts to opponents that appear larger and nearer. These frequency alter-

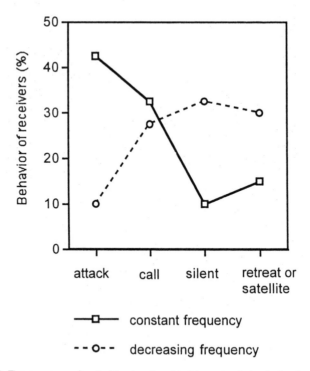

FIGURE 4.10. The response of male Blanchard's cricket frogs to playback of calls that either maintained a constant frequency or decreased in frequency by 200 Hz, from Wagner (1992). Calls of decreasing frequency were more successful than calls of constant frequency in inducing receivers to cease calling (silent) or to retreat or adopt satellite behavior. (In satellite behavior the frog ceases calling and adopts a low posture in the water.) Calls of constant frequency were more likely to induce an attack on the speaker.

ations are not trivial; Wagner (1989) found that dominant frequency can change as much as 360 Hz, representing approximately a third of his study population's range of variation. Furthermore, receivers respond to frequency shifts in a way that should be beneficial to signalers; male cricket frogs were more likely to abandon calling and retreat, and less likely to attack, when exposed to playback of calls that decreased in frequency than when call frequency was constant (figure 4.10; Wagner 1992).

One interpretation of the cricket frog results is that calling males alter their dominant frequencies in order to deceive opponents, causing opponents to assess caller size as greater than it actually is (Wagner 1989, 1992). One problem with this deception hypothesis is that, if callers always make a predictable frequency alteration in response to a certain aggressive stimulus, then receivers could evolve different assessment rules to be used in different aggressive contexts to compensate for these alterations. Consequently, a given dominant frequency would be taken as representing one size for an undisturbed caller and a second, smaller size for a male calling after an aggressive challenge, and size assessment would be equally accurate under either condition. As might be expected, however, the degree of frequency shift in response to a given aggressive challenge actually varies among males (Wagner 1989). Furthermore, dominant frequency is a better predictor of body size for males calling undisturbed ($r^2 = 0.45$) than for males calling at the end of playback ($r^2 = 0.18$) (Wagner 1992).

These results accord with the deceptive use of dominant frequency by cricket frogs, but Wagner (1992) considers two alternative hypotheses. One is that the size of the frequency shift itself signals body size. This hypothesis is easily dismissed, for both Wagner (1992) and Burmeister et al. (2002) found essentially no correlation between body size and the magnitude of frequency shift in response to playback. A second alternative is that the size of the frequency shift is a signal of aggressive intentions. In accord with this hypothesis, Wagner (1992) found a significant correlation between the magnitude of the frequency shift made by males in response to playback and the probability that they would attack the speaker: the greater the shift the greater the probability of attack. Burmeister et al. (2002) later confirmed that a drop in dominant frequency predicts attack. Wagner (1992) argues that just the opposite would be expected under the deception hypothesis: males would exaggerate their size more for superior opponents than for inferior ones, and so males giving a large frequency shift would be less likely to attack than those giving a small shift or none.

Wagner's results suggest the following scenario for the evolution of signaling via dominant frequency in Blanchard's cricket frogs. Initially, dominant frequency is correlated with the body size of calling males for anatomical reasons. Because of this correlation, receivers evolve to withdraw from low-frequency calls. This effect on receivers is advantageous to signalers, especially when challenged, so signalers then evolve the tactic of lowering dominant frequency in response to an aggressive challenge by another male. At this point, frequency shifts are correlated with aggressive behavior by signalers, in the sense that the shift is a reliable signal that a caller is aware of, and is responding to, a challenge. Receivers then evolve to respond to this reliable signal of the caller's aggressiveness, that is, to the frequency shift.

If this scenario is correct, can we say that frequency shifts are deceptive? A strong case can be made that shifts are deceptive midway through the hypothetical evolutionary trajectory, at the point where the tactic has initially appeared in the population. The case is more ambiguous at the endpoint. Here, if a frequency shift has become a reliable signal of aggressiveness, can it at the same time be considered to be deceptive about size? We would answer in the affirmative, because of the evidence that receivers attend to absolute frequency as well as to frequency shifts, and because of the evidence that frequency is a poorer predictor of size after the shift than before.

Given that calling at low dominant frequencies has demonstrable benefits in Blanchard's cricket frogs, the fact that undisturbed males are capable of shifting to lower frequencies when challenged implies that there must be some cost to low-frequency calling; otherwise, males would be expected never to call at the higher frequency. One possible cost is diminished amplitude (Bee et al. 2000). Bee and Perrill (1996) found that in aggressive contexts male green frogs decrease both the dominant frequency and the amplitude of their calls. If males control dominant frequency by varying pulmonary air pressure, then lowering dominant frequency would result in a decrease in amplitude as a biomechanical by-product (Martin 1971). A decrease in calling amplitude might very well impose a fitness cost, especially in terms of attracting fewer females.

In green frogs, unlike Blanchard's cricket frogs, dominant frequency is just as well correlated with body size after dominant frequency has shifted in response to an aggressive challenge as it is before such a shift (Bee et al. 2000). In theory, then, receivers should be able to assess the size of the caller just as accurately after the shift as before, if they are able to compensate for the context of calling. Whether receivers are capable of such compensation is unknown. Frequency shifts have been found in a few other species of anurans, but none of these cases provides promising candidates for deception. In American toads (*Bufo americanus*), males lower dominant frequency when interacting with other males, but dominant frequency is a better predictor of body size for interacting males than for noninteracting ones (Howard and Young 1998). Carpenter frogs possess two frequency peaks in their calls; of the two, only the primary peak is correlated with body size, whereas only the secondary peak shifts in response to challenges (Given 1999). In white-lipped frogs (*Leptodactylus albilabris*), males are able to alter the dominant frequency of their chirp calls in response to playback, but are as likely to increase frequency as to decrease it (Lopez et al. 1988). The effect of these alterations is to cause the subject's calls to converge on the frequency of the playback stimulus. This kind of "matching" may be analogous to the matching of song types in birds, which has been suggested to serve as an unambiguous signal that the signaler doing the matching is attending to the caller being matched.

In conclusion, the anatomy of the vocal signaling structures in anurans acts to produce a correlation between dominant frequency and body size, simply because large males tend to have large vocal cords and large cords produce low frequencies. Nevertheless, the ability of some frogs to alter their dominant frequencies in response to context demonstrates that the signaling system is not constrained to be honest in any absolute sense. The best case that frequency shifts actually are used to deceive listeners can be made for Blanchard's cricket frog, in which downward shifts are effective in intimidating opponents, and in which the shifted calls given during aggressive contexts are less accurate predictors of body size than are calls given in undisturbed calling.

VOCAL CUES TO BODY SIZE IN MAMMALS

The argument that call frequency is constrained to be an honest indicator of body size is often made for animals other than anurans. Again, the length of the vocal cords is expected to be of primary importance in determining the minimum fundamental frequency of a vocalization (Fitch and Hauser 2003). If this assumption is met, and if the length of the vocal cords is constrained to be tightly related to body size, then minimum frequency will be constrained to reveal body size. We have already seen, however, that the assumption that fundamental frequency will be correlated with the length of the vocal cords is not met in toads (Martin 1971). Furthermore, Fitch and Hauser (2003) have argued that in mammals the size of the larynx, and hence the length of the vocal folds, has considerable flexibility in an evolutionary sense, and therefore is not constrained to a tight correlation with body size. As examples of the evasion of this supposed constraint, they cite the increase in the size of the larynx in human males relative to human females that occurs at puberty, and the hypertrophy of the larynx in male hammerhead bats (*Hypsignathus monstrosus*), in which the larynx is three times that of the female and fills the entire thoracic cavity. Because the size of the larynx can vary independently of body size, a relationship between body size and fundamental frequency is not inevitable in mammals, and in fact no such relationship has been found when it has been looked for, whether in humans (van Dommelen and Moxness 1995), red deer (*Cervus elaphus*) (Reby and McComb 2003), or Japanese macaques (*Macaca fuscata*) (Masataka 1994).

A second possible vocal cue to body size is maximum call length (Fitch and Hauser 2003). Maximum call length may in part be determined by the volume of air that can be exhaled without drawing breath, and thus by lung volume. Lung volume in turn may be tightly correlated with body size. If these two assumptions are met, then maximum call length would be constrained to be an honest indicator of size. Constraints on call length, however, can be evaded in ways that parallel those we have seen with call frequency: anatomically, for example through the evolution of elastic air sacs, which increase the volume of air that can be passed over the vocal cords without an increase in lung volume, and behaviorally, through the trade-off between call length and call intensity (Fitch and Hauser 2003).

A more promising possibility for a constrained acoustic cue to body size is formant dispersion (Fitch 1997). "Formants" are the frequencies that are emphasized by the resonances of the vocal tract in humans. Fitch (1997) has suggested that the term be applied to animal vocalizations in general, even though the sounds produced by nonhuman animals differ significantly from those produced by humans. "Formant dispersion" refers to the average spacing between successive formants. If the vocal tract approximates a simple tube, then the spacing between formants should be inversely correlated with vocal-tract length, and if vocal-tract length is constrained by body size, then body size and formant dispersion should be negatively correlated (Fitch 1997). Fitch and Hauser (2003) have argued that, for mammals, the relationship between body size and vocal-tract length is more difficult to evade than the one between larynx size and body size, because the components of the vocal tract (the pharyngeal, oral, and nasal cavities) have so many functions beside vocal production, making them less free to vary for signaling considerations. Fitch and Hauser (2003), however, point out a number of general ways that the relationship between body size and vocal-tract length can be evaded. Morphologically, the vocal tract can be lengthened at either end, by lowering the larynx in the throat, as occurs in human males at puberty, or by evolving an elongated proboscis, as in male proboscis monkeys (*Nasalis larvatus*). Behaviorally, the vocal tract can be lengthened by using muscles to pull the larynx lower when vocalizing, by raising the head, and by closing the mouth and pursing the lips.

Empirically, formant dispersion has been found to be strongly correlated with body size in rhesus macaques (*Macaca mulatta*), with an r^2 of 0.78 (Fitch 1997), and in domestic dogs (*Canis familiaris*), with an r^2 of 0.77 (Riede and Fitch 1999). In humans, however, body size and formant dispersion have no relationship (Collins 2000). Red deer provide a particularly interesting example, because they are known to use behavioral mechanisms for manipulating the length of their vocal tracts when roaring, by lowering the larynx and raising the head (Fitch and Reby 2001). Maynard Smith and Harper (2003) argue that since all red deer stretch their vocal tract maximally when they roar, formant dispersion is still a reliable signal of size, constrained to honesty by the relationship between maximal vocal-tract length and body size. In fact, however, Reby and McComb (2003) report that red deer do not always maximally extend their vocal tracts during roaring, perhaps because doing so requires considerable effort. Reby and McComb (2003) analyzed the first roars in bouts, since these tended to show minimum formant dispersions and by inference maximum vocal-tract extension, and found that formant spacing was related to body size with an r^2 of 0.39. Formant dispersion thus seems to be roughly as good a predictor of body size in red deer as dominant frequency is in anurans. If all roars, rather than just first roars, were analyzed, presumably the reliability of this signal property would be lower.

Conclusions

In all four of the aggressive signaling systems we have reviewed, receivers have been shown to respond to the signals in question. The evidence is particularly compelling for badges of status in birds and dominant frequency in frog calls, for in both cases experimental manipulations have demonstrated the effectiveness of the signals, that is, there is direct experimental evidence that the signals themselves trigger a response in receivers. In other systems we have experimental evidence of response but not of effectiveness (e.g., for open-chela displays in snapping shrimp) or correlational evidence that production of the display is followed by a receiver response beneficial to the signaler (e.g., avian postural displays and the meral-spread display of stomatopods). Although the quality of the evidence varies, there is little reason to doubt the existence of receiver response in any of these systems.

Furthermore, in all these systems the signals have been shown to contain information important to receivers in assessing opponents. Avian postural displays predict the next act of the signaler with some accuracy, though this may not be their primary message. Badges of status are strongly correlated with age, sex, and (sometimes) size, all often important in determining fighting ability. Weapon displays in crustaceans predict aggressive escalation (in intermolt stomatopods) or body size (in snapping shrimp). Dominant frequency in frog calls in general is correlated with the caller's body size, a trait very important to fighting ability. None of these signals is absolutely reliable, but as we have argued before, it is hard to imagine any animal signal that would appear to us as absolutely reliable.

Given the relative reliability of these aggressive signals, it is easy to understand why receivers have evolved to respond to them. What is not so apparent is why the signals are as reliable as they are. Some of the signals we have discussed fit well with the handicap idea, in the sense that they impose a cost that falls more heavily on individuals of poorer quality, but in others the applicability of the handicap principle is not obvious. Badges of status provide two cases in point. In Harris's sparrows, aggression is especially common between individuals of similar badge size, so that when cheaters are created by artificially enhancing their badges, they encounter increased aggression from particularly dominant (and thus dangerous) individuals—a near-perfect example of a receiver-dependent cost. In white-crowned sparrows, by contrast, aggression is not especially common between individuals of similar badge size, and artificially created cheaters do not experience increased aggression from anybody. In this species, then, there is no receiver-dependent cost, and yet badges seem to be just as reliable as in Harris's sparrows. Exaggerated badges in white-crowned sparrows may have a cost that has not yet been identified, but this just points to the conclusion that we do not know enough about the cost side of the equation in many of these signaling systems.

The robust but imperfect reliability of these aggressive signals leaves ample room for deception on the part of some individuals, and indeed there is good evidence for deception in some but not all of these systems. The meral-spread display of stomatopods provides an especially clear case of deception, clear in that we can unambiguously identify deceivers (newly molted individuals giving the display), and hence observe the consequences of their deceptions. More typical, however, may be cases where deceptive individuals differ from honest ones continuously rather than discretely, as in the open-chela display of snapping shrimp. In systems such as this, deceptive individuals can only be identified by first determining the general relationship between the signal and the signaled attribute (in this case between chela size and body size) and then locating those individuals that depart from the relationship in a particular direction (larger chela than expected). Dominant frequency in frog calls provides a parallel case, though here short-term alterations of the signal ease identification of possible deceivers in species like Blanchard's cricket frog. In neither snapping shrimp nor cricket frogs is it certain how receivers interpret exaggerated signals, but in both cases there is intriguing evidence that signalers benefit from exaggeration. Further study of "signal residuals" in species with continuously varying signals of fighting ability may prove central to our attempts to understand reliability and deceit in animal communication.

5 Honesty and Deception in Communication Networks

Our analysis of honesty and deception thus far has taken as its starting point an implied view of communication as a fundamentally dyadic interaction, with a sender and a receiver that may differ in their evolutionary interests, but which nonetheless interact with each other independently of the influences of other actors. This dyadic view of communication has provided an appropriate platform for our discussion, for two reasons. First, viewing communication as a dyadic interaction has long been the dominant perspective in studies of animal communication (Marler and Hamilton 1966, Brown 1975, Wilson 1975), and most of the literature has taken this perspective as its starting point. Second, the dichotomy between sender and receiver defines the evolutionary conflict of interest in a signaling interaction and thus captures the essential issue underlying reliability and deceit. Nevertheless, it is commonly—perhaps almost invariably—the case that a signal will be detected by more than one receiver. McGregor and his colleagues have introduced the idea of "communication networks" to describe the broader social environment in which signaling may occur (McGregor 1993, McGregor and Dabelsteen 1996, McGregor and Peake 2000, McGregor et al. 2000).

The costs and benefits of a signaling interaction may not be altered substantially by adding extra receivers to the mix, if all of those receivers fall into the same functional class. When several female crickets detect the mate-attraction call of a single male, or when both parent birds detect the begging calls of chicks in a nest, the analysis of costs and benefits may be more complicated, because there are additional individuals to consider, but the fundamental equation does not necessarily change. In other cases, however, the different potential receivers of a signal may introduce different selective pressures. Signals produced in an interaction between two territorial males, for example, are likely to be detected by other males not involved in the interaction, as well as by females (McGregor 1993, McGregor et al. 2000). The exchange between the two signaling males may help to determine the outcome of the aggressive interaction between them, and the honesty of signaling will be enforced by a receiver-dependent cost (see chapter 4). But what of the other potential receivers? Both females and other males not directly involved in the dyadic interaction may gain information about the two interacting individuals, and change their future behavioral responses to one or both males on the basis of that information. The potential costs and benefits to both the signaler and the receiver in the dyadic interaction must include the effects of the signaling interaction on these other, "third-party" receivers in the network. The additional costs and benefits so produced may affect the mix of reliability and deceit found in the signaling system at evolutionary equilibrium.

Third-Party Receivers

The idea that individuals of other species can listen in on and exploit signaling exchanges among conspecifics is long established (Marler 1955, Otte 1974, Myrberg 1981). Indeed, much early debate in the animal-communication literature centered on the question of whether a signal intercepted by an unintended receiver of a different species constitutes "true" communication (e.g., Marler 1977, reviewed in Bradbury and Vehrencamp 1998). Is it communication, for example, when a gecko finds its prey by orienting to the calling of male crickets (Sakaluk and Belwood 1984)? Or when a parasitoid fly locates an ant host by following its alarm pheromones (Feener et al. 1996)? Whether or not the exploitation of signals by natural enemies is considered communication, third-party receivers such as these clearly may impose serious costs on both the signaler and the intended receiver in a signaling interaction.

We have previously considered detection by predators as a potential cost for signals such as the begging of nestling birds and the carotenoid coloration of guppies. Another, particularly well-described example is provided by the mate-attraction call of the túngara frog (*Physalaemus pustulosus*). Male túngara frogs produce a complex call beginning with a frequency-modulated "whine" that may be followed by one or more harmonically rich "chucks." The number of chucks produced by a male increases with his body size, and females respond more to calls having a greater number of chucks, suggesting that the number of chucks is an honest indicator of size used by females in mate choice (Rand and Ryan 1981, Ryan 1983). The honesty of this signaling system would be ensured if the chucks were energetically costly to produce and the larger males were better able to support the cost of production. This mechanism appears not to operate, however. Calling by túngara frogs is energetically costly (Ryan 1985a), but most of the cost stems from production of the invariant whine component of the call; the added chucks do not significantly increase the overall cost (Ryan 1985b). Instead, the major cost of producing more chucks is that incurred from predators: fringe-lipped bats (*Trachops cirrhosus*) home in on the calls of túngara frogs and are more likely to prey on frogs producing calls having several chucks than on frogs producing only whines (Ryan et al. 1982). Thus, the potential for an unintended receiver in this case imposes the cost that most significantly affects the signaling system.

Interception of signals by conspecific individuals outside the primary sig-naler/receiver dyad may also affect the evolution of a signaling system. Inter-ception by conspecifics typically will result in a less drastic outcome for the signaler than being eaten, but may nonetheless have a significant impact on the costs and benefits of signaling to both signaler and receiver. Two possible ef-fects may be pertinent to signal reliability. First, the response of third-party receivers may impose additional costs on signaling that could conceivably be important in ensuring reliability. As an example, a signal feature might be par-ticularly effective in attracting potential mates while also attracting same-sex competitors; only males that can bear the costs of aggressive interaction with competitors would find it profitable to produce the signal. Effects in this first category are directly analogous to those stemming from the interception of signals by predators, although again the cost of same-species interception might be lower than the cost of interception by predators. The second possibility is that third-party receivers might observe both the signal and the subsequent interaction between signaler and receiver, and use their memory of that interac-tion to shape their own response when the same signaler signals to them at some later time. As an example, a food call given by a rooster to one hen might be observed by a second hen, who also observes whether the first hen obtains food, and uses these observations in shaping her own subsequent responses to that male. Here the effect of interception by third-party conspecifics is an exten-sion of the individually directed skepticism we discussed in chapter 2.

"Eavesdropping" versus "Interception"

In many cases a third-party receiver can make use of a signal to ascertain some-thing about the signaler, perhaps something as simple as its location, without reference to the interaction in which the signaler is engaged. In such cases, the signaling interaction itself is not directly relevant to determining the added costs of an unintended receiver, other than to have provided the context for the production of the signal in the first place. A less well-studied consequence of signaling in a communication network arises when individuals observing a signaling interaction obtain information about the participants, not from the signal itself, but from the interaction in which that signal is used. McGregor and his colleagues have dubbed this latter case "eavesdropping" (McGregor 1993, McGregor and Dabelsteen 1996, McGregor and Peake 2000).

McGregor and Dabelsteen (1996, p. 416) specify that it is "a prerequisite of eavesdropping that a third party (the eavesdropper) gains information from an interaction that could not be gained from a signal alone." Eavesdropping thus refers specifically to the acquisition of information about the relative qualities of individuals, such as their relative resource-holding potential, dominance, condition, motivational state, and so forth, all obtained from observing a signaling interaction. Eavesdropping can be distinguished from "information gathering" (McGregor and Dabelsteen 1996) or "interception" (Myrberg 1981), two terms used to refer to the more general case in which third-party receivers make use of information passed between the primary signaler and receiver in a dyadic interaction. Eavesdropping is thus a subset of interception, that subset where the outside receiver gains information from the interaction rather than just from the signal. "Eavesdropping," however, is not always used in the literature in as restrictive a fashion as suggested by McGregor and colleagues, instead being used as another synonym for interception. This relaxed usage is understandable, since in everyday English eavesdropping can be applied equally well in the broad sense (any interception of signals by third-party receivers) as in the narrow. Peake (in press) has suggested using the terms "interceptive eavesdropping" and "social eavesdropping" to acknowledge the common usage of the term "eavesdropping" while at the same time maintaining the distinction between intercepting signals in general versus attending to signals in the context of the outcome of a social interaction. We agree with McGregor, Peake, and their colleagues that the distinction is a meaningful one in animal communication, although we will use the established terms "interception" and "eavesdropping" in our discussion, the latter used only in the narrow sense originally proposed by McGregor and Dabelsteen (1996), as the extraction from a signaling interaction of information that cannot be obtained from the signals alone.

To clarify the distinction between eavesdropping and interception, consider a territorial male songbird as a member of a communication network that includes other territorial males and their mates in the same neighborhood. This individual is the intended receiver of many of the signals it encounters, of course, such as the broadcast songs that serve to delineate the territories of its neighbors. At the same time, the male may be the unintended receiver of other signals produced in the network, such as male courtship signals directed specifically toward females (e.g., Balsby and Dabelsteen 2002) or aggressive countersinging between two males contesting a particular boundary (e.g., Langemann et al. 2000). The former case would represent an example of interception, in the sense that our third-party male generally will be able to monitor only the signals of one of the interactants (the singing male) because the other (the female) either signals at very low amplitude or does not signal at all. Here, interception of the signal may inform the third-party male of the presence of a receptive female on the neighbor's territory, affecting the likelihood he will intrude. Interception of countersinging provides a more complicated example. Overhearing a countersinging interaction would be simply a case of interception if the third male learns that one of the contestants is in good condition, on the basis of its high song rate, or is of good phenotypic quality, judged from its high vocal performance. Overhearing the contest would be a case of eavesdropping if our third male is able to learn that one of the two males he overhears interacting is in better condition than the other, on the basis of their relative song rates, or that one is dominant over the other, from the temporal relationship of their signals or from direct observation of the outcome of a fight. Information gained from either "eavesdropping" or "interception" may be used by the male later as it makes behavioral choices, for example into which neighboring territory to trespass in search of extra-pair copulations.

Although we will seek to maintain the distinction between eavesdropping and interception, we at the same time acknowledge that both processes can influence the costs and benefits of honest signaling and thus the balance between reliability and deceit. We next review the role of eavesdropping in the strict sense in animal communication. We then consider the implications of both eavesdropping and interception for the reliability of animal signals.

Eavesdropping in Signaling Interactions

To explore the evolutionary stability of eavesdropping in aggressive contests, Johnstone (2001) developed an extension of Maynard Smith's (1982) classic Hawk-Dove game, which includes, in addition to hawks and doves, a third "eavesdropping" strategy. In this model, hawks are assumed to employ their usual strategy of all-out aggression, and doves their usual strategy of nonaggression. Eavesdroppers employ a strategy that is contingent on the previous success or failure of their opponent, playing hawk if their opponent has lost its previous encounter and playing dove if their opponent has previously won. The usual payoffs are obtained for winning and losing, and the value of a victory (v) is assumed to be less than the cost of a defeat (c), which is the necessary condition for a mixed equilibrium in the standard Hawk-Dove game. Under these conditions, Johnstone (2001) found stable equilibrium frequencies that include a combination of hawks, doves, and eavesdroppers, the relative proportions of these strategies depending on the ratio v/c, again as in the standard Hawk-Dove game. Thus, Johnstone (2001) was able to demonstrate that eavesdropping can be a stable strategy in aggressive-signaling contests, replacing some proportion of doves in the population with a strategy that depends on information gathered from observing the interactions of others.

This much is not surprising, given that one frequently cited advantage of eavesdropping in aggressive-signaling systems is that it allows the eavesdropper to assess opponents before interacting with them, and thus avoid contests it is likely to lose. Johnstone's (2001) model, however, also leads to two less-intuitive conclusions. The first is that the eavesdropping strategy does not spread to fixation in a population, in spite of the advantage eavesdroppers have in using information from the interactions of others, at no cost to themselves. The reason for this, according to Johnstone, is that eavesdroppers benefit only if they can anticipate predictable behavior on the part of their opponents. This advantage is lost when one eavesdropper encounters another, because the eavesdropping opponent's behavior is contingent on its previous observations, and so is not as predictable as that of a hawk or dove. As the eavesdropping strategy increases in a population, the predictability on which this strategy depends is increasingly lost, and the strategy becomes less advantageous.

A second, somewhat paradoxical result of Johnstone's (2001) model is that the presence of the eavesdropping strategy increases rather than decreases the overall level of escalated contests predicted to occur in the population, as compared to what happens in a population having only hawks and doves. This result is paradoxical because one would intuitively assume that escalated contests should decrease, given that eavesdroppers will avoid escalating contests with known hawks. The presence of eavesdroppers, however, provides an added benefit to winning a contest because the winner of one contest is more likely to win the next, given the possibility that its next opponent may be an eavesdropper that is less likely to escalate, having observed the previous win. In this way, eavesdropping adds to v, the value of a victory, and in so doing leads to increased escalation overall (Johnstone 2001). This theoretical result, then, suggests that the presence of eavesdroppers may increase receiver-dependent costs of aggressive signaling in an unexpected way, by increasing the number of aggressive challenges.

Terry and Lachlan (in press) have argued that existing game-theory and population-genetics models, although useful for studying communication in dyadic interactions, are relatively ill-suited for modeling communication in networks, because they typically assume that interactions occur randomly in a population, and that animals have perfect knowledge of the behavior of all other individuals with which they might interact. Terry and Lachlan suggest that individually based, spatially explicit simulations may be more useful for modeling the effects of eavesdropping on signal evolution. Preliminary work using this approach, for example, suggests that eavesdropping may be a more prevalent strategy than Johnstone's (2001) model allows when eavesdroppers observe and interact with only those individuals that are located nearby, rather than with individuals that are drawn randomly from the population at large (Terry and Lachan in press). The use of individually based, spatially explicit simulations for modeling eavesdropping (and other aspects of communication in networks) seems promising, but this work is only beginning, and it remains to be seen how such models will contribute to theory in this area.

EVIDENCE FOR EAVESDROPPING IN AGGRESSIVE SIGNALING

Recent empirical studies on aggressive signaling in birds and fish have lent considerable support to the idea that individuals gain information in a communication network from eavesdropping. A fundamental way to demonstrate that eavesdropping has occurred is to allow one individual to observe a signaling interaction between two other individuals and then ask whether the observer's behavior toward either of these individuals changes in a way that is consistent with the information it could have gained from eavesdropping on their interaction. An early experiment of this kind was done by Freeman (1987), who used taxidermic mounts to elicit aggressive responses from territorial red-winged blackbirds and then recorded subsequent intrusions (by neighbors of the test subjects) across established territory boundaries. The idea behind this experiment was that neighbors might eavesdrop on the staged interaction and change their behavior toward the test subject, depending on how aggressively it interacted with the simulated intruder. As one might expect, not all subjects in Freeman's tests attacked the mount with equal vigor, and some did not attack at all. Subsequently, neighbors were more likely to trespass on the territories of test subjects who failed to attack the mount, which is consistent with the hypothesis that neighbors had learned through eavesdropping that these individuals were less aggressive toward intruders.

There is an alternative explanation for Freeman's (1987) result, however, which underscores the difficulty in demonstrating eavesdropping in a signaling system. Males that failed to attack the mount in Freeman's tests may have been less aggressive in general, perhaps because they valued their territories less or because they had low fighting ability, compared to the males that attacked the mount. The tendency of neighbors to trespass on the territories of the low responders might then be explained by the neighbors' own direct experience in interacting with these nonaggressive males, rather than by their having observed the interaction between the nonaggressive males and the test mount. It follows that to support the interpretation that eavesdropping has occurred it is necessary to show that any change in the behavior of a potential eavesdropper not only is consistent with information it might have gained from observing an earlier interaction, but also depends on the eavesdropper's actually having observed that specific interaction.

Several experiments satisfying this additional requirement have been done with fish in aquarium settings, where it is relatively easy to monitor and control the experience of potential eavesdroppers. For example, Johnsson and Åkerman (1998), working with rainbow trout (*Oncorhynchus mykiss*), housed two individuals together on one side of a partitioned aquarium while a third individual (the observer) was housed on the other side. Fish on opposite sides of the tank could see and smell each other, but could not make physical contact. The two individuals housed together interacted aggressively, one soon becoming dominant over the other. After a dominance relationship had been established, Johnsson and Åkerman then paired the observer fish either with the "familiar" dominant fish (i.e., the individual that had won the contest occurring earlier on the opposite side of its tank) or with an "unfamiliar" dominant fish (i.e., an individual that had won a contest, but in a different tank, so that the contest was not seen by the observer). Whether or not the observer was paired with a

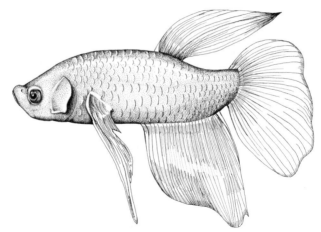

FIGURE 5.1. A Siamese fighting fish.

familiar or an unfamiliar fish did not affect the outcome of its own contest with that fish—there was an equal probability of the observer winning or losing, as would be expected for size-matched fish. The dynamics of this subsequent contest, however, did appear to be influenced by eavesdropping. Observers that were paired with familiar dominant fish settled their contests (either winning or losing them) more quickly and with less aggression than observers paired with unfamiliar dominant fish. The interpretation here is that the information obtained by the observers eavesdropping on the earlier contests allowed them to decide more rapidly whether or not to challenge their opponents.

Oliveira et al. (1998) performed a similar experiment with Siamese fighting fish (*Betta splendens*) (figure 5.1) in which the experience of contestants and potential eavesdroppers was controlled even more completely. Again, contests were staged between two fish, with a third fish allowed to observe the interaction. Because Siamese fighting fish will fight to the death if housed together, the contestants in this case were separated by a glass barrier and could interact only visually. In this circumstance, one fish eventually will display less and assume a more submissive posture, and the amount of time spent displaying during a trial can thus be used as a proxy for winning the contest (Simpson 1968, Evans 1985, Peake and McGregor 2004). The observer was separated from the two interacting fish by one-way mirrored glass, eliminating the possibility of signaling between the observer and the interacting fish. When observers were paired subsequently with fish they had watched interacting, they took a significantly longer time to approach those individuals they had seen win than to approach individuals they had seen lose, and they displayed longer to winners than to losers. By contrast, there was no difference in either measure when individuals interacted with winners and losers of contests they did not observe, which, again, is consistent with an eavesdropping interpretation.

In both the Johnsson and Åkerman (1998) and Oliveira et al. (1998) studies, individuals were paired with winners (or losers) of contests that they had either observed or not observed. This design not only tests whether an observer must see the contest for its behavior to change, it also addresses another alternative explanation for the results: that differences in the outcome of interactions between observers and opponents they have seen in an earlier contest are not due to eavesdropping, but instead are due to changes in the opponents' behavior as a result of having previously won or lost. Such "winner effects" and "loser effects" are known to occur across a broad range of taxa (Jackson 1991, Chase et al. 1994, Hsu and Wolf 1999, 2001). Simply put, an animal that has won one contest is more likely to win the next all other things being equal. Because the effects of winning or losing a previous contest were effectively held constant in both the Johnsson and Åkerman (1998) and Oliveira et al. (1998) studies, this interpretation is ruled out. More recent studies on green swordtails (*Xiphophorus helleri*) by Earley and Dugatkin (2002) reached a similar conclusion.

In the results we have reviewed so far, it is possible in each instance that the behavior of an observer changed after seeing an interaction not because it has observed the interaction per se, but more simply because it has had the chance to observe something intrinsic about the participants that it would not have seen otherwise. In other words, the results could all be explained more simply by signal interception rather than by eavesdropping. The difference here is subtle, but important for the definition of eavesdropping proposed by McGregor and his colleagues, which emphasizes the ability of an observer to extract information about the relative qualities of two individuals specifically from the signaling interaction between them. Consider the fighting fish example. The amount of time that an individual spends giving a particular aggressive display, such as gill-cover erection, might intrinsically reflect something about the quality of the signaler, as for any condition-dependent display. Having observed an opponent in a previous contest, an observer may modify its behavior toward that opponent simply on the basis of its earlier direct assessment of that signal, irrespective of the relative signaling properties of the interactants or the outcome of the contest it observed. The only effect of the signaling interaction, by this interpretation, is to create a context that makes those traits apparent (Peake and McGregor 2004). To demonstrate eavesdropping in the strict sense one must eliminate this alternative.

One way to test for eavesdropping in the strict sense is to determine whether the behavior of an observer toward two interactants is better predicted by the interactants' relative behavior (e.g., who displays more) than by their absolute behavior (e.g., their absolute rates of display). McGregor et al. (2001) carried out such a test using fighting fish. One male, the observer, was allowed to watch two other males through one-way glass. In one treatment, the two males interacted aggressively with each other; in the second, the two males appeared to be interacting with each other but were actually interacting with two other males hidden from the observer's view (figure 5.2a). Winners of both real and apparent interactions were designated on the basis of which fish spent more time in the aggressive gill cover erect posture close to the opponent. When the observers were subsequently exposed to each interactant in turn, they were more aggressive toward winners than losers, at least in the apparent interaction treatment (figure 5.2b, c). Absolute levels of aggressive behavior of the interactants did not correlate with differences in the observers' responses to the winner and loser.

An even more elegant approach to controlling for the absolute characteristics of signalers in an interaction is to simulate both of the participants of an observed contest, thus allowing the experimenter to hold constant the intrinsic signal properties of the apparent contestants and to manipulate only the information available from the signaling interaction itself. Song-playback experiments with birds have provided a convenient system for conducting this kind of experiment. Male birds often vary the timing and delivery of songs relative to each other during aggressive interactions, and in some cases these patterns are thought to convey information about aggressive intent or relative dominance of the interacting birds (Todt 1981, McGregor et al. 1992, Dabelsteen et al. 1996, 1997, Todt and Naguib 2000). Naguib and Todt (1997) used this phenomenon to design an eavesdropping experiment in which a territorial male nightingale is confronted by two simulated intruders, the intruders being identical except for an asymmetry in their signaling interaction.

When nightingales (*Luscinia megarhynchos*) are countersinging, one male may begin to sing before the song of the other male has finished, thus "overlapping" its opponent's song (Hultsch and Todt 1982). Males avoid returning to song posts at which they have been overlapped (Todt 1981) and adjust the timing of their songs to avoid being overlapped (Hultsch and Todt 1982), suggesting that being overlapped is aversive (Todt and Naguib 2000). Naguib and Todt (1997) placed a pair of loudspeakers a few meters inside a male nightingale's territory and about 15 meters apart. Songs played from either speaker provoked an aggressive response from the territorial male, as would occur in a conventional-playback experiment. If songs are played from both speakers, the subject might be expected to respond more to one speaker than to the other, presumably depending on its assessment of the relative threat of the two simulated intruders. By offsetting the timing of songs between the two speakers, Naguib and Todt could make one of the simulated intruders overlap the other during playback.

Naguib and Todt (1997) found that test subjects responded more strongly to the speaker playing the overlapping song than they did to the speaker from which the overlapped songs were played, as measured by the amount of time birds spent and number of songs they sang in closer proximity to one speaker

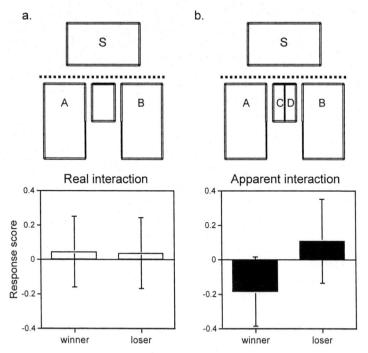

FIGURE 5.2. An experimental demonstration that Siamese fighting fish respond to relative levels of signaling behavior in an aggressive interaction, as required by the strict definition of eavesdropping (from McGregor et al. 2001). Four aquaria are arranged so that the subject ("S") can observe two other fish ("A" & "B") that have either a real interaction with each other or an apparent interaction. Between the aquaria housing fish A and fish B is a smaller aquarium that may or may not contain fish, depending on the treatment. Dark lines indicate opaque barriers, and dotted lines indicate a one-way mirror allowing fish S to observe A and B, but not vice versa. Response scores represent the first principal component derived from an analysis of eight variables related to aggressive signaling (such as time spent with erect gill covers, number of tail beats, etc.) that explains 58.5% of the variance in the data. The middle aquarium is empty in one treatment (**a**, top), allowing A and B to see and interact with each other. In a second treatment (**b**, top), the middle aquarium has a partition and contains two additional fish ("C" & "D"). In this case, A and B appear to be interacting with each other, but in fact are interacting with C and D, respectively. Whereas observers generally respond more to observed winners than to losers, they responded significantly differently to opponents depending on whether they had seen them in real or apparent interactions, with a greater difference in aggressive response (indicated by a more negative PC1 score in this analysis) between apparent winners and losers (**b**, bottom) than between real winners and losers (**a**, bottom). The interpretation here is that this difference results from a mismatch in the actual outcome of the original interaction as compared to the apparent outcome seen by the observer. In all cases, absolute measures of size, size differences, or display characteristics of fish A or fish B did not predict the response of fish S to either, whereas the outcome of their real or apparent interaction did.

than to the other. Naguib and Todt interpreted this result as suggesting that subjects perceived the simulated overlapping bird to be a greater threat, and therefore the more important intruder to confront. Mennill and Ratcliffe (2004) obtained the same result with black-capped chickadees (*Poecile atricapillus*), using an experimental design very similar to that of Naguib and Todt (1997), although the outcome depended on the dominance status of the subject, with only high-ranking males showing a clear tendency to approach the speaker playing the overlapping song. Clearly, in both of these studies, information obtained by observing the interaction (or listening to it, to be more precise) influenced how the observer subsequently behaved toward the simulated contestants. More important, because the songs played from either speaker could be considered functionally equivalent in all ways other than their relative timing, the only information available to the observer that could distinguish between the two interacting individuals must have come from the signaling interaction itself, not from anything intrinsic to the signals or signalers.

One weakness with this interpretation is that subjects may have responded more to the overlapping signal simply because the overlapping signal was the last signal they heard. This idea does not necessarily undermine the conclusion that eavesdropping has occurred, but by offering a simple proximate mechanism based on increased attention to the most recent signal heard, it does suggest a less rich framework for interpreting how observers extract information about the relative qualities of individuals. Naguib et al. (1999) addressed this issue in a follow-up study using the same playback design with nightingales, except that the second song played immediately followed the first but did not overlap it. In this leader-follower arrangement, they argued, the preceding song now represents the more aggressive or dominant individual because the follower delays its song to avoid overlapping. Consistent with this interpretation, birds in this experiment responded more strongly to the song heard first (the leader). Although there is no independent evidence for the function of leader-follower roles in countersinging by nightingales, the result nonetheless shows that the response of the eavesdropper to an asymmetry in the vocal interaction cannot be explained by a simple attention mechanism.

Working with great tits, Peake et al. (2001) expanded on the design introduced by Naguib and his colleagues, increasing the ecological validity of the experimental design and controlling variation in the signaling interaction more completely. As in other species, patterns of overlapping during countersinging are thought to signal relative aggressiveness in great tits, the overlapper being more likely to escalate an encounter (McGregor et al. 1992, Dabelsteen et al. 1996, Langemann et al. 2000). Furthermore, great tits appear able to recognize individuals on the basis of idiosyncratic voice characteristics, independent of the particular song types they sing (Weary and Krebs 1992, Lind et al. 1996), making it possible to separate in time the staging of an interaction between two countersinging males and the staging of a later interaction between an

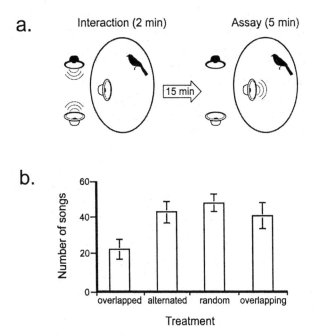

FIGURE 5.3. A demonstration of eavesdropping in great tits using song playback (from Peake et al. 2001). **a.** Schematic illustrating the playback design. A territorial male great tit hears a staged interaction played through two loudspeakers located outside his territory for 2 minutes. Fifteen minutes later, one of the simulated interactants is made to "intrude" by playing its songs through a third loudspeaker located on the subject's territory during a 5-minute assay. **b.** Response of the territorial male to the staged intrusion depends on the nature of the interaction it overheard. The subject responds less (in terms of number of songs produced) to perceived losers, which are birds whose song had been overlapped in the prior interaction. The subject responds equally strongly to perceived winners (those simulated intruders who were overlapping in the overheard interaction) and to intruders about which it has ambiguous information (with alternating or randomly spaced songs in the overheard interaction).

eavesdropper of that contest and one of the contestants it overheard. Like Naguib and his colleagues, Peake et al. (2001) began their experiment with an "interaction playback" in which they simulated a countersinging interaction using two loudspeakers, the songs differing in the timing of their delivery relative to each other. Unlike what was done in the work with nightingales, however, both loudspeakers were located 5 to 10 meters *outside* the test subject's territory, and thus neither of the simulated males was an immediate threat during the eavesdropped interaction (figure 5.3a). Fifteen minutes later, during the "assay playback," songs from one of the males simulated in the interaction playback were broadcast from a third loudspeaker placed inside the subject's territory, mimicking an intrusion by one of the overheard contestants, and the response of the subject to this intrusion was measured. This arrangement more realistically approximates the context in which a territorial male might make use of information it has gained by eavesdropping (by contrast, it is unlikely that a male would encounter a simultaneous invasion by two males who were also having an aggressive contest with each other, as modeled by the nightingale experiments). In addition to simulating interactions in which one bird overlapped another during the interaction playback, Peake and his colleagues simulated interactions in which songs from the two speakers were played in alternation or with completely random timing with respect to each other.

As predicted, the behavior of male great tits in assay playbacks depended on the interaction they heard earlier, although the nature of their response differed from that observed in nightingales. Males sang significantly less when confronted with the songs of a bird they had previously heard being overlapped (figure 5.3b). When subjects were confronted with a bird they had heard overlapping, however, they sang at the same rate as when they were confronted with a bird they previously heard alternating songs or singing songs with random timing. That is, males responded less to an observed loser of a contest, but responded the same way to observed winners and to "ambiguous" males. Peake et al. (2001) interpreted these results to mean that information gained from eavesdropping is useful only when interacting with a known loser, which unambiguously poses a lesser threat; males for which information is unclear (those that had sung in alternation with, or randomly, with respect to another bird) evoke the same level of response as known winners because all might be serious threats.

A different kind of evidence demonstrating the occurrence of eavesdropping comes from studies of aggressive signaling in primates. Dorothy Cheney, Robert Seyfarth, and their colleagues (Cheney et al. 1995, Bergman et al. 2003) have asked whether baboons (*Papio cynocephalus ursinus*) recognize an anomalous aggressive interaction in which a dominant individual reacts submissively to an individual known to be of lower rank in the hierarchy, something that rarely occurs under normal circumstances. As with the bird studies described above, an interaction between two individuals was staged using playback, and the response of a third individual overhearing this interaction was observed. Baboons hearing an aggressive signal (a "threat grunt") produced by a low-ranking individual followed by a submissive signal (a "scream" or "fear bark") produced by an individual of higher rank attended to the interaction significantly more, as measured by time spent gazing in the direction of the signals, than when they heard a dominant individual's threat grunt followed by a lower-ranking individual's scream (Cheney et al. 1995, Bergman et al. 2003). Irrespective of the functional significance of this response difference, it is clear that the third-party observer attended to information derived from the signaling interaction that it could not have obtained by hearing either signal on its own, which is consistent with the definition of eavesdropping in the narrow sense.

In the real world, eavesdroppers are likely to have interacted previously with individuals they observe in contests with others, providing them directly with information about the competitive abilities of these individuals, information that may be more accurate than, or perhaps even contradictory to, information they glean from eavesdropping. How are these different sources of information integrated? Earley and Dugatkin (2002) examined this question in their study of green swordtails. An observer fish was separated from a contest between two other males by an opaque partition, a clear partition, or a one-way mirror. With the clear partition, contestants could interact with the observer (albeit only through visual signaling), thus directly providing the observer with information about their aggressive abilities. When later matched against the winner of the first contest, the observer was less likely to defeat the winner if it had been separated from the first contest by a mirror than if it had been separated by either the clear or the opaque partitions. One way to think of these results is that the opaque condition allows winner/loser effects only, the mirror condition allows winner/loser effects and eavesdropping, and the clear condition allows winner/loser effects, eavesdropping, and direct interaction. Thus, the fact that the observer's chance of winning is lower in the mirror condition than in the opaque condition implies an effect of eavesdropping, whereas the fact that the observer's chance of winning is no better in the clear treatment than in the opaque implies that the effect of eavesdropping is cancelled by direct interaction.

Peake et al. (2002) examined the integration of eavesdropping and direct information in greater detail by modifying their playback protocol with great tits to include a prior territorial intrusion by one of the simulated contestants later overheard by the eavesdropper. The initial intrusion was simulated using playback from a single loudspeaker inside the subject's territory (figure 5.4a). By controlling the timing of songs in this playback relative to the songs produced by the test subject—either overlapping the subject's songs or allowing the subject's songs to overlap the playback—Peake and his colleagues could establish the simulated intruder as either the "winner" or the "loser" of this initial interaction with the subject. They then arranged for the simulated intruder of this first playback to interact outside the territory with a second simulated bird, as before, the second bird either winning or losing, again on the basis of the timing of song delivery. In the final assay playback, the second simulated bird was now made to appear to intrude on the subject's territory, and the response of the subject to this intruder was measured. As in their earlier study, male great tits reduced their song output in response to intrusion by an individual they had overheard losing a contest (figure 5.4b). This effect was only observed, however, if this intruder had lost to a bird that previously had lost to the subject in the initial intrusion playback. If the second intruder had

a.

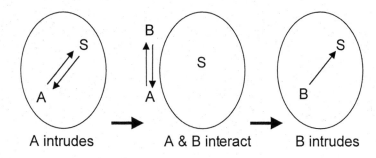

A intrudes   A & B interact   B intrudes

b.

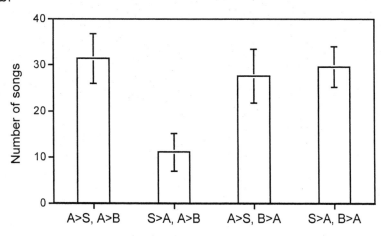

FIGURE 5.4. Integration by male great tits of information gained in eavesdropping with information gained in direct experience (from Peake et al. 2002). **a.** Schematic illustrating playback design. A territorial subject ("S") first interacts with a simulated intruder ("A"), being allowed either to "win" or "lose" the interaction (depending on whether it is allowed to overlap or caused to be overlapped by the experimenters). The subject next hears bird A interacting with a second simulated bird ("B") outside its territory. Finally, bird B is made to appear to intrude on the subject's territory. **b.** Response of subject males (in terms of number of songs sung) to intrusion by bird B. The response is significantly less only in the case where S had won its interaction with A, and A had won its interaction with B (S > A, A > B), which is consistent with an interpretation that S had gained information about the relative aggressiveness of A and B by eavesdropping. The strong response to B in the treatment in which B loses to A after S loses to A indicates that information obtained through eavesdropping is combined with the eavesdropper's own experience. If information obtained through eavesdropping alone was used, the response would be expected to be weak because B had been observed to lose. However, the fact that S also lost to A makes this information ambiguous.

lost to a bird that had apparently won against the subject, then the subject responded as aggressively as it would have to a perceived winner (figure 5.4b). Peake and his colleagues suggested that in the latter case the relative competitive abilities of the second intruder and subject are ambiguous, because both had apparently lost to the same third party. Whether or not this interpretation is correct, this result nonetheless shows that male great tits combine information obtained by eavesdropping with their own direct experience.

EVIDENCE FOR EAVESDROPPING IN MATE CHOICE

To this point, we have discussed the effects of eavesdropping only on the outcome of aggressive interactions. Empirical studies show that females, too, change their behavior toward individuals depending on the outcome of interactions they have observed previously. If signals used in aggressive interactions between males are reliable about traits such as condition or phenotypic quality, then females too might well benefit from eavesdropping on such interactions. Otter et al. (1999) provided the first demonstration of such an effect by staging interactions between territorial male great tits and simulated intruders, and then observing the behavior of the females to which these males were mated. During the staged interactions, the intruder playback was made either to "de-escalate" the encounter by alternating its songs with those of the territory holder or to "escalate" by overlapping the territory holder's songs. Pairs of neighbors were tested, one neighbor being randomly chosen to receive the de-escalated treatment and the other to receive the escalated treatment. The assumption of this design is that males interacting with a submissive, de-escalating playback would appear to be having an easier time evicting the intruder than would males interacting with an aggressive, escalating playback.

Otter et al. (1999) found that females mated to subjects receiving the escalated treatment left their mate's territory and intruded on a neighbor's territory significantly more often than females mated to subjects receiving the de-escalated treatment (figure 5.5). In most cases, females leaving their social mate's territory intruded on the territory of the neighbor they had heard apparently winning an interaction with an intruder more easily. Whether or not a female left her mate's territory and intruded on another's could not be predicted by any absolute measures of singing, but was predicted by the singing behavior of her mate or neighbor relative to the playback treatment they received (figure 5.6), leading Otter and his colleagues to conclude that female behavior in this case had been influenced by eavesdropping.

One interpretation of these results is that females paired with males experiencing aggressive, escalated playbacks perceive their social mates as being of lower quality as a result of eavesdropping, and thus are more prone to seek extra-pair fertilizations from neighbors, especially neighbors appearing to be of higher quality because they interacted with a de-escalating intruder. An

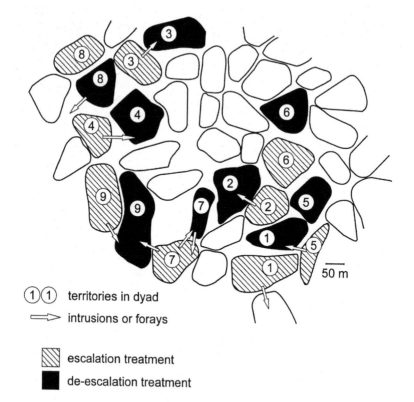

(1)(1) territories in dyad

⟹ intrusions or forays

▨ escalation treatment

■ de-escalation treatment

FIGURE 5.5. Map of territories showing the movements of female great tits following staged playback interactions with territorial males (from Otter et al. 1999). Territories in which playbacks occurred are shaded, with hatching indicating escalating playbacks that made the territorial male appear to have a relatively lower song output, and darker shading indicating de-escalating playbacks having the reverse effect. Territories were tested in matched pairs, which are numbered. The arrows show observed female movements following playbacks. Seven of nine females mated to males given the escalation treatment intruded on neighbors' territories, six of these onto territories of males that had been given the de-escalation treatment. Only one of nine females mated to males given the de-escalation treatment intruded onto a neighbor's territory (another female, #8, left her territory but went to an unoccupied area). (Reprinted with permission from K. Otter et al. 1999. Do female great tits [*Parus major*] assess males by eavesdropping? A field study using interactive song playback. *Proceedings of the Royal Society B* 266:1305–1309. Published by The Royal Society.)

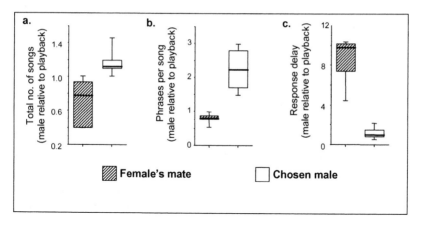

FIGURE 5.6. Differences between "chosen" males and a female's social mate in song characteristics relative to the playback treatment they received (from Otter et al. 1999). **a.** Number of songs. **b.** Number of phrases per song. **c.** Response delay to opponent's song. These data are taken from the six females that were mated to escalation-treatment males and that intruded onto the territory of other males in the study. For all measures, the song characteristics of "chosen" males relative to their playback treatment were significantly superior to those of the social mates relative to the playback they encountered, which is consistent with the view that female behavior was influenced by eavesdropping on the playback interactions. (Reprinted with permission from K. Otter et al. 1999. Do female great tits [*Parus major*] assess males by eavesdropping? A field study using interactive song playback. *Proceedings of the Royal Society B* 266:1305–1309. Published by The Royal Society.)

analysis of genetic parentage by the individuals in this study, however, failed to find any evidence that playback manipulations influenced female choice of extra-pair mates (Otter et al. 2001). Otter and his colleagues suggested that the interactions they staged were too short-term to outweigh other sources of information available to females, and thus insufficient to cause a detectable change in mate-choice decisions.

A subsequent study of black-capped chickadees did find a measurable effect of eavesdropping on extra-pair mate choice. Mennill et al. (2002, 2003) followed the same procedure as Otter and his colleagues, although they additionally assigned males as being either high-ranking or low-ranking on the basis of flock interactions the preceding winter. Previous work had demonstrated that the great majority of lost paternity in this species is suffered by low-ranking males, whereas high-ranking males typically lose very little paternity (Otter et al. 1998). When confronted with an aggressive, escalating playback, however, high-ranking males lost paternity at a significantly higher level than high-ranking males receiving a de-escalating playback or no playback at all. The paternity loss of low-ranking males was unaffected by playback, remaining high even if they experienced the more submissive, de-escalating treatment (Mennill et al. 2002). The mates of high-ranking males receiving an aggressive playback treatment mostly obtained extra-pair fertilizations from other high-ranking males, as is typical for this species; in those few cases in which females obtained extra-pair fertilizations from low-ranking males, they did so from males that had received a submissive playback treatment and thus had apparently won an aggressive contest recently (Mennill et al. 2003). These findings suggest, for black-capped chickadees at least, that eavesdropping by females on even a short-term interaction between males can have a direct effect on the fitness of the participants of that interaction.

Laboratory experiments with Siamese fighting fish and Japanese quail (*Coturnix japonica*) also have demonstrated an effect of eavesdropping on female mating behavior. Doutrelant and McGregor (2000) showed that female fighting fish spend more time associating with, and display reproductive color more often to, the winners of staged contests they observe than to the losers. Females showed no such preferences if they were permitted to observe males immediately after a signaling contest but were not permitted to observe the contest itself or its outcome, demonstrating that differences in female behavior must be due to eavesdropping on the interaction and not to winner/loser effects on male behavior or coloration. Ophir and Galef (2003) used a conceptually similar design to ask whether observing an aggressive contest between two males influenced the subsequent affiliative behavior of Japanese quail females. In this case, female eavesdroppers affiliated more closely with the loser than with the winner of the contest they observed. The reason for this reversed preference, Ophir and Galef (2003) suggest, is that more aggressive males are more likely to injure females during courtship. Exactly how the affiliative behaviors measured by Doutrelant and McGregor (2000) and Ophir and Galef (2003) would translate into differences in mating success is unclear, but their results, taken together with the work of Otter et al. (1999) and Mennill et al. (2002, 2003), illustrate how females may eavesdrop on contests between males as a way of obtaining information about the characteristics of potential mates.

INTERCEPTION, EAVESDROPPING, AND THE AUDIENCE EFFECT

Because signal interception and eavesdropping have the potential to change the relative costs and benefits of a signaling interaction, we would expect the participants in an interaction to change their behavior depending on whether or not they are being watched and on who is watching. "Audience effects" of this sort have been studied in the context of food calling by chickens, where males are more likely to call upon finding a food item if a female is present than when alone (Marler et al. 1986b, Evans and Marler 1994). Work on Siamese fighting fish also has demonstrated that the presence of an audience can affect the dynamics of an aggressive contest, changing the proportions of different signals used, and further that these changes depend on the sex of the audience.

In the presence of a female audience, male fighting fish interacting with another male reduced the proportion of highly aggressive displays, increased the proportion of highly conspicuous displays, and spent more time in close proximity to their opponent, as compared to contests where no observers were present (Doutrelant et al. 2001). In the presence of a male audience, interacting fighting fish escalated faster, attempted more bites, and spent less time near their opponent (Matos and McGregor 2002), although some of this effect could be attributed to aggressive priming due to the presence of the observer before the contest began (Matos et al. 2003). Matos and McGregor (2002) argue that the sex-specific audience effects they observed are consistent with expected costs and benefits of eavesdropping. On the one hand, females may avoid highly aggressive males, owing to the risk of being bitten (Bronstein 1984), so using particularly aggressive displays in the presence of a female observer may have an added cost. On the other hand, a third-party male observing another male losing a contest or simply behaving less aggressively may be more likely to challenge that male (e.g., Freeman 1987, Oliveira et al. 1998), thus increasing the benefit of showing strong aggression in the presence of a male observer (Matos and McGregor 2002). It is worth noting here that it is difficult to ascribe the effects observed by Matos and McGregor to eavesdropping, strictly defined, because the same predictions hold if the third-party observers are merely intercepting signals and not evaluating the relative signaling in the contest or its outcome.

A compelling example of how individuals in a signaling interaction might mitigate the added costs imposed by interception or eavesdropping comes from a study by Herb et al. (2003), who compared the subsequent behavior of male fighting fish involved in a signaling contest toward females that had observed that contest versus "naive" females that had not. Contest losers subsequently spent significantly less time displaying toward a female that had seen them lose than they did toward a naive female (figure 5.7). By contrast, winners showed no preference between an eavesdropper and a naive female. These investigators interpreted this result in light of Doutrelant and McGregor's (2000) suggestion that females are less likely to mate with a fish they have seen lose a contest—by avoiding courtship with females that have seen them lose, males are able to minimize this cost of eavesdropping. In this example, the effects are easier to attribute to eavesdropping than to interception because the change in male behavior was contingent on the outcome of their own contest, not just on the presence of the female observer.

CONCLUSIONS

Taken together, studies of fish and birds provide substantial evidence that eavesdroppers can gain information about the aggressive capabilities or moti-

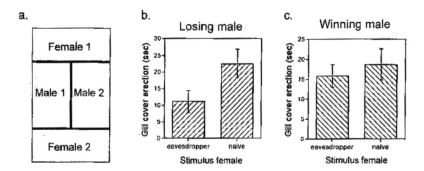

FIGURE 5.7. The response of male Siamese fighting fish to females that have either seen or not seen them interact with another male, from Herb et al. (2003). **a**. Schematic diagram illustrating the experimental aquarium setup, in which one or the other of two female fighting fish could observe an interaction between two males. Opaque partitions could be introduced to isolate tanks visually from each other. One female was allowed to observe the males interact until one male had clearly won the contest, while the other female was prevented from doing so. Males could observe females as well. Following the contest, the males were visually isolated from each other, and the responses of winning and losing males to females that had either seen or not seen the contest were measured, by adding and removing opaque partitions as appropriate. **b**. Responses of losing males to a female that had observed them lose an interaction ("eavesdropper") and to a female that had not observed the interaction ("naive"), measured by the number of gill-cover erections, a typical sexual display. Losing males responded significantly more to naive females. **c**. Responses of winning males to eavesdropping and naive females; here there is no significant difference in male response.

vation of individuals they observe in interactions with others, although the direction of the observed effects differs from system to system. Nightingales signal more to winners (Naguib and Todt 1997). Siamese fighting fish were shown to be slower to display to males they observed win an aggressive contest in one study (Oliveira et al. 1998) but faster and more vigorous in their display to a winner in another (McGregor et al. 2001). In green swordtails, eavesdropping affects only the outcome of contests with observed winners, not observed losers (Earley and Dugatkin 2002), whereas great tits decrease their signaling to observed losers, signaling the same to winners as they do to individuals about which they have only ambiguous information (Peake et al. 2001). Clearly, interpreting the functional consequences of eavesdropping remains a challenge, but having demonstrated that eavesdropping as well as signal interception may be a common occurrence in some signaling systems, we now turn to the question of how this phenomenon may affect signal reliability.

Third-Party Receivers and Reliability

As we argued above, one way in which signal interception can affect signal reliability is for interception to produce a signaling cost that falls more heavily on some classes of signalers than on others. Differential costs can then make reliable signaling the optimal behavior for all classes of signaler, as envisioned by the handicap principle. Candolin's (1999, 2000) work with sticklebacks, discussed in chapter 3, furnishes one example of how costs may be imposed by third-party receivers. Recall that, in this example, the extent of red carotenoid coloration exhibited by males decreased when males were held in groups, as compared to when they were held in isolation. The decrease of red in the presence of conspecific males suggests that, in addition to physiological or other costs associated with this signal, there also is a receiver-dependent cost imposed not by the presumed primary recipient of the signal (the female) but instead by other males intercepting the signal and reacting aggressively to it. The correlation between the area of red exhibited by males and male parental ability was significantly stronger for males with decreased red following inter-actions with other males, suggesting that the receiver-dependent cost imposed by third-party interceptors plays an especially important role in maintaining the reliability of this signaling system.

Perhaps the best example of a case in which third-party receivers impose differential costs that maintain signal reliability involves signaling of domi-nance in male brown-headed cowbirds. Female cowbirds tested in isolation from males give the copulation-solicitation posture in response to playback of male songs (King and West 1977). Song appears to be a necessary stimulus for female copulatory behavior, for females do not solicit at all unless a male has sung to them (West et al. 1981). Female cowbirds held in groups copulate preferentially with the dominant male in their group. Female preference for dominant males seems to be based at least in part on song, since females are more responsive to the songs of dominants than to songs of subordinates when songs are played to them in the absence of males (West et al. 1981). Females are also more responsive to songs of males that have been reared in isolation than to those of males with normal experience (King and West 1977), and placing males in isolation after they have started to sing leads to an increase in the effectiveness of their songs in eliciting female display (West and King 1980). The particular "potency" of the songs of isolates and dominants appears to be related to the "interphrase unit" or IPU, a short (50-msec), high-frequency phrase that has an especially high relative amplitude in isolate songs (West et al. 1979). Playback of songs in which the amplitude of the IPU has been artifi-cially lowered results in a diminished female response, as compared to play-back of unmanipulated songs (West et al. 1979). The implication of these results is that the IPU is a marker of dominance status, and that female cowbirds use this signal to identify and mate with dominant males.

Given that producing songs with loud IPU's is of benefit to males in stimulating a sexual response in females, we can ask how the reliability of this signal is maintained. In other words, why do not all males produce loud IPU's, regardless of their true dominance status? Unlike the example of red coloration in sticklebacks, there seems no obvious cost to producing the signal in the first place. It is conceivable that there is a performance constraint operating, such that only high-quality males are capable of producing a loud IPU. This hypothesis is undermined, however, by the fact that males held in isolation are able to increase the potency of their songs (West and King 1980). Instead, the reliability of the IPU as a signal of dominance status appears to be maintained solely by the response of other male cowbirds. When introduced to a group, males that have been isolated and are singing high-potency songs are attacked and even killed by males already in the group (West and King 1980). Dominant males, when moved from one group to another, also are attacked (West et al. 1981), whereas males that have been subordinate in one group, and which therefore sing low-potency songs, are not harmed when moved between groups (West and King 1980). Thus the reliability of a male's signal to a female is maintained by the response of third-party receivers of the same species. In a stable group, the cost ought to fall differentially on males according to their quality; that is, subordinate males would pay a higher cost for giving the signal of dominance than would truly dominant males. Altogether, this seems an excellent case of the maintenance of signal reliability by the response of third-party receivers.

In both of the examples just cited, third-party receivers are responsible for imposing the cost that enforces signal reliability, illustrating the potential importance of considering signaling systems in the context of a communication network, as opposed to a dyadic interaction, in order to understand their evolutionary stability. In neither example, however, is there evidence that eavesdropping, as opposed to simple signal interception, is at play. Further, in both examples, it is not clear that eavesdropping, if it were occurring, would significantly change the nature of the cost imposed by third-party receivers. The finding that female great tits (Otter et al. 1999) and black-capped chickadees (Mennill et al. 2002) change their behavior toward males, depending on the outcome of contests they observe, may represent a case in which a novel cost (possible loss of paternity) resulting specifically from eavesdropping is introduced into a signaling system, but in these cases it is unclear whether a potential for deceit exists in the first place. That is, the reliability of aggressive signaling between males in these cases is likely to be maintained by receiver-dependent costs imposed by the competing male in a dyadic interaction with no need to invoke the cost imposed by female eavesdroppers to explain signal reliability.

The fact that we can find no clear example of eavesdropping, strictly defined, affecting the reliability of a signaling system is not surprising, given that convincing demonstrations that eavesdropping occurs at all have only recently appeared. We are left, then, to ask in theory how eavesdropping could affect the stability of a signaling system, beyond imposing novel costs or benefits. One possibility is a mechanism that extends individually directed skepticism to the context of eavesdropping. In our previous discussion of individually directed skepticism (see chapter 2), we suggested that receivers remember the reliability of signals produced by particular individuals in past interactions and use that information to shape their responses when they receive signals from the same individuals in subsequent interactions (Silk et al. 2000). We cited evidence from a number of studies that receivers do remember the past reliability of signals received from particular individuals, and that these memories do affect subsequent responses. Given this evidence, and the evidence that eavesdropping occurs, it seems reasonable to posit that individuals might gain information about the reliability of individual signalers through eavesdropping, as well as through their own direct interactions. Thus an eavesdropper might observe an aggressive interaction between two other individuals, A and B, in the course of which A gives an aggressive signal that is or is not followed by an attack. Our assumption here is that the signal normally is a reliable indicator of impending aggression, but sometimes may be used deceptively without an actual attack. The eavesdropper could use its memory of these events when it interacts with A, responding to A's aggressive signals if those signals had been honest in the eavesdropped interaction, and ignoring them if they had not been honest. This mechanism of "third-party skepticism" would provide an additional incentive for reliability, in the sense that deception carries the cost of diminished future response, not only on the part of the receiver but also on the part of any other individuals that happened to have eavesdropped on the deceptive interaction.

We know of no models or empirical studies that have explored third-party skepticism, but the mechanism is analogous to "indirect reciprocity" through "image scoring" as modeled by Nowak and Sigmund (1998). In the Nowak and Sigmund model, altruistic acts of A toward B are favored even if B can never reciprocate, because other individuals observe A's altruism and this enhances A's image (or reputation) as an altruist. Other individuals are more likely to help A if A has built up a positive image, and less likely to help A if A's image is negative. Nowak and Sigmund's analysis shows that systems of indirect reciprocity can be made evolutionarily stable through this image-enhancement mechanism. In the Nowak and Sigmund model the long-term benefits of maintaining a positive reputation for altruism offset the loss of immediate benefits that could be accrued through selfishness. By analogy, it seems reasonable that the long-term benefits of maintaining a reputation for signal reliability might lead to forgoing the immediate benefits of deception.

Not surprisingly, Nowak and Sigmund find that cooperation through indirect reciprocity is easier to establish in small populations, because in a smaller population a larger proportion of the population is able to observe any given interaction. Similarly, we would expect that reliability should be easier to maintain through third-party skepticism in small groups than in large.

Conclusions

It is apparent that animals often, perhaps usually, communicate in networks of signalers and receivers, not simply in dyadic pairs. There is ample evidence demonstrating that third-party receivers intercept signals, and demonstrating how interception may affect the costs and benefits of a signaling interaction to both signalers and receivers. McGregor and his colleagues have proposed that communication networks create the opportunity for an even more specialized kind of signal interception referred to as "eavesdropping," in which observers of an interaction gain information from the relative signaling of the interactants or from the outcome of the interaction than they would from intercepting the signals themselves (McGregor 1993, McGregor and Dabelsteen 1996, McGregor and Peake 2000). Recent work, especially studies of aggressive signaling in fish and birds, demonstrates that animals can gain information from eavesdropping in this fashion.

It remains unclear, however, how much a network view of communication may change our fundamental understanding of signal reliability and deceit. In at least two examples, those involving carotenoid signaling in sticklebacks (Candolin 1999, 2000) and song in cowbirds (West et al. 1979, 1981, West and King 1980), there appears to be good evidence that third-party receivers in a communication network are responsible for imposing the cost that is primarily involved in maintaining signal reliability. These examples can be viewed as rather simple extensions of the receiver-dependent cost idea, wherein reliability is maintained by the responses of third parties to the signal rather than by the response of the primary receiver. Third-party skepticism provides a more radical proposal for how eavesdropping could influence signal reliability. One could subsume third-party skepticism under the handicap principle, with the consequences of lowering one's reputation for reliability as just another handicap cost, but we would argue that the idea of maintaining one's image of reliability is sufficiently different to justify separating the two hypotheses. Note that third-party skepticism requires eavesdropping in the strict sense; the third party must observe not only the signal but some details of the interaction, such as whether the signal is followed by aggression. As of now we know of no empirical support for third-party skepticism, but this seems a fruitful area for future research.

6

Conclusions

From what we know about how natural selection works, we can assume that animals will produce signals only if doing so increases their own fitness. Similarly, we can assume that receivers will respond to signals only if doing so increases *their* fitness. The sole value of a signal to a receiver is as a source of information, information that it uses in choosing the behavioral, physiological, or developmental responses that will maximize its fitness. The set of responses that is best for the receiver, however, will only rarely be identical with the set that is best for the signaler. Selection therefore should often favor transmission by the signaler of information that is misleading, in the sense that it induces the receiver to respond in a way that increases the signaler's fitness rather than its own. If the information in signals is misleading often enough, however, receivers will be selected to ignore the signals. And if receivers ignore signals, signalers do not benefit from giving them, and we expect the signaling system to disappear. This, in short, is the honest-signaling dilemma. To explain why signaling persists—that is, to solve the dilemma—we need to identify mechanisms that maintain the reliability of signals produced by signalers that are pursuing their own selfish interests.

The above account, an account that has framed our discussion throughout this book, relies heavily on the concept of information. That concept has had a checkered history in the animal communication literature. At one point, communication was defined in terms of information transfer (Marler 1968, Batteau 1968), and empirical studies sought to produce quantitative measures of the amount of information contained in various animal displays (Hazlett and Bossert 1965, Dingle 1972). Dawkins and Krebs (1978), however, argued that the expectation that any information is transferred at all is based on the implicitly group-selectionist assumption that communication is inherently cooperative. Rejecting that assumption, and viewing the act of signaling as essentially manipulative, Dawkins and Krebs (1978) suggested that it was more useful "to avoid the very idea of information." The criticisms of group-selectionist approaches to animal communication, raised by Dawkins and Krebs as well as others, such as Otte (1974), were tremendously influential in shaping current views on animal signaling. Nevertheless, the admonition to reject the idea of information never took hold, and the concept remains central to the field. The basic reason is given in the argument above: the only value of a signal to a receiver is as a source of information, and without this value the receivers will not respond to the signal and the signalers will not produce it. An individual-selection analysis of animal communication, when taken to its logical conclusion, thus predicts that reliable information will be contained in signals often enough to make them "honest on average."

Reliability

In the interval since individual-selectionist thinking came to dominate studies of animal communication, much of the theory in the field has been devoted to explaining how the information in a signal can be reliable enough to support the evolutionary maintenance of receiver response. We have reviewed major segments of that theory in this book. A pivotal idea for much of this theory has been Zahavi's handicap principle, the idea that signals can be honest if they are costly in some appropriate way. Grafen's (1990a,b) models provided the most general exposition of this principle, and these models have been especially influential in convincing others of the validity of the idea that reliable signaling depends on signal cost. The usual way of interpreting these models is that embodied in figure 1.1a: costs and benefits increase with signaling level, the costs increasing more rapidly for low-quality signalers than for high-quality signalers, so that optimal signaling levels increase with signaler quality. Most of the more specific models we have reviewed include this kind of cost set-up as an essential assumption, particularly models of mating and aggressive signaling.

Taken together, signaling models lead to a series of empirical expectations, or predictions, about the nature of animal signaling systems. These predictions are: (1) that receivers will respond to signals, (2) that signals are reliable enough to justify receiver response, and (3) that signals are costly in a way that explains why they are reliable. In the preceding chapters we have reviewed evidence for these predictions from a variety of types of signaling systems; below, we briefly summarize this evidence as a prelude to considering how well theory is doing.

PREDICTION 1: RECEIVERS WILL RESPOND TO SIGNALS

That receivers respond to signals is not surprising; what would be surprising is if animals were selected to give signals that receivers ignore. But the animal kingdom is filled with examples of unique and conspicuous movements, markings, smells, and sounds, many of which may have nothing to do with communication. Only by demonstrating that receivers respond to a particular behavior or image can one prove that a signaling system exists. For this reason, we started the analysis of each signaling system by examining the evidence for receiver response. In general, the evidence for receiver response is quite strong. Some of this evidence is correlational, but often it has been possible to perform experiments demonstrating that a receiver responds to a signal in a manner consistent with the signal's presumed function. So, for example, parental feeding rates correlate with offspring begging rates, and when begging signals are experimentally manipulated, parents respond by feeding more. For carotenoid mating signals, female preferences correlate with natural variation in the amount and intensity of red colors in males, and females can be induced to change their preferences by artificially enhancing those colors. Similarly, for tail length in birds with elongated tails, and for aspects of bird song such as rate, complexity, and performance, female preferences correlate with the display trait and can be altered by manipulating the display.

Status signals provide one of the most compelling cases for receiver response, compelling because it is not obvious a priori that receivers pay any attention at all to these signals. Natural variation in status signals such as the dark breast plumage of Harris's sparrows and the bold crown markings of white-crowned sparrows correlates with aggressive dominance. Much of the variation in these plumage characters can be explained by age and sex, traits that themselves often affect dominance. It seems entirely possible that the correlation between dominance and plumage is no more than an indirect consequence of changes in plumage with age and sex that are adaptive in some other context, or perhaps not adaptive at all. Nevertheless, when these signals are artificially manipulated, individuals move up or down in dominance status as expected, demonstrating that the signals must have an effect on receivers.

PREDICTION 2: SIGNALS ARE RELIABLE ENOUGH TO
JUSTIFY RECEIVER RESPONSE

As we have noted more than once, testing this prediction is complicated by the problem of deciding what it is that the signal is supposed to be reliable about. This is not a trivial problem. Sometimes it is possible to demonstrate that a signal affects female choice of mates, or the outcome of aggressive encounters, without our being able to determine what the information is that is conveyed by the signal responsible for its effect. The best we can do is to focus on information that receivers would benefit from knowing in a given signaling context. Ideally, we would like to determine whether the signal is reliable enough that the receiver's fitness is, on average, increased by attending to the signal; in practice, we have to be satisfied with finding out whether the signal has any reliability at all.

Some of the best evidence on reliability comes from studies of begging. For parents, as receivers, information about their offspring's need ought to be of benefit both in allocating time and effort to feeding young versus other activities, and in allocating food among their current young. Various aspects of begging behavior have been shown to increase in intensity with food deprivation and to decrease with feeding, demonstrating that begging is reliable about short-term need. Begging intensity in some cases is also responsive to past food deprivation, demonstrating that the signal contains information about long-term need as well. Other signals given between relatives also show considerable reliability. Alarm calls in chickens and vervet monkeys, for example, generally are reliable about the presence of some threat, although they are less reliable about the specific nature of the threat. Food calls in birds and primates tend to be reliable about both the presence and the quality of food.

Reliability may not seem surprising in signals given to relatives, but reliability in signals given to potential mates is another matter. Selection ought to favor exaggerating one's quality to potential mates whenever possible. In spite of this expectation, available evidence suggests that mating signals can be highly reliable, in the sense of conveying information about signaler quality that receivers benefit from knowing. Carotenoid pigmentation in fish and birds varies with the quality of a male's diet and his exposure to parasites, and can be used to predict the quality of his future parental care and perhaps his genetic quality as well. Song rates in birds reflect a male's diet and may predict the quality of his territory and future parental care. Song complexity and other aspects of song structure are affected by a male's early developmental history, and therefore reflect his overall phenotypic quality. Tail length in barn swallows and peafowl reliably conveys information on the quality of male genes affecting survival, genes that are important to the fitness of a potential mate's young.

Reliability seems even less of a given in aggressive signaling, since in this context the interests of signaler and receivers are usually in direct opposition. Measuring reliability acquires an extra layer of complexity whenever a signal conveys intention rather than ability or quality; this extra complexity arises because the future behavior of the signaler may be contingent on the receiver's response to the signal, and so is not perfectly predictable, even by the signaler itself, at the time the signal is given. Nevertheless, at least some aggressive signals appear to be reliable about aggressive intentions. A particularly interesting case is provided by avian postural displays, the category of signal used by Caryl (1979) and others to question the reliability of aggressive signals in general. Postural displays are informative about future aggressive behavior in species such as blue tits (Stokes 1962) and American goldfinches (Popp 1987), whether or not receiver response is taken into consideration. Other categories of aggressive signals are reliably correlated with attributes of the signaler that affect its fighting ability; good examples are provided by chela displays in snapping shrimp (Hughes 1996), badges of status in Harris's sparrows (Rohwer et al. 1981), and dominant frequency of advertising calls in various anurans.

PREDICTION 3: SIGNALS ARE COSTLY IN A WAY THAT
EXPLAINS WHY THEY ARE RELIABLE

Costs have been adduced for virtually every signal we have discussed. Suggested costs come in a range of forms, including increased expenditure of energy and time, depressed immunocompetence, increased vulnerability to predation, impairment of flight performance, decreased investment in the development of other aspects of the phenotype, and increased aggression from receivers. Despite the evidence that exists for these various costs, the empirical support for the prediction that signals are costly in a way that explains their reliability has been sharply questioned (Kilner and Johnstone 1997, Kotiaho 2001), and our analysis reinforces this critical view in a number of cases.

One problem with the evidence for signal costs is that the magnitude of the costs, once this is measured, often turns out to be lower than expected. Expectations here are most often derived intuitively, since theory usually does not offer quantitative predictions in a testable form on the magnitude of costs needed to stabilize a signaling system. Begging provides a good case in point (Kilner and Johnstone 1997). Two types of costs, energy expenditure and predation risk, have been suggested for begging in nestling birds. The rate of energy expenditure does increase during begging (McCarty 1996, Leech and Leonard 1996). The amount of increase is small relative to what is seen in older birds when they are exercising—though it may be close to the maximal increase that can be achieved by very young birds (Chappell and Bachman 1998). Nevertheless, the increase is small enough, and begging is infrequent enough, that the cost as a proportion of the total energy budget seems negligible. Begging has been shown to have negative effects on growth, but only in a minority of the species that have been studied. As for the other suggested cost, experimental evidence from playback of begging calls at nests shows that begging increases the risk of predation appreciably for nests located on or near the ground (Haskell 1994, Leech and Leonard 1997, Dearborn 1999), but not for nests placed higher up (Haskell 1994), and there is no evidence to suggest that begging is less reliable in birds that nest in trees than it is in ground-nesters. Furthermore, certain specific begging signals, such as the red mouth color of canaries, seem particularly unlikely to be costly, and yet even these are reliable. Thus, evidence that begging is costly in a way that could maintain reliability is mixed at best.

Another criticism that has been leveled at the evidence on signaling costs is that even when some cost can be shown, that cost is usually not the kind of direct fitness cost that is assumed by models of honest signaling (Kotiaho 2001). To demonstrate a direct fitness cost, one needs to show that a signal lowers either the survival or the reproduction of the signaler. A few studies of signal costs do meet this criterion, notably in cases where a signal increases predation risk, as with carotenoid pigmentation in fish. More typically, however, empirical studies demonstrate a cost that is several steps removed from fitness, such as increased expenditure of energy or time or decreased immunocompetence. Kotiaho (2001) argues that demonstrations of what he refers to as "indirect costs" are inadequate as evidence for honest-signaling models, because the evidence rests on the unproven assumption that such indirect costs lead to fitness differences. Kotiaho (2001) makes this argument with specific reference to sexual signals, but the same criticism can be applied to existing evidence on costs in other types of signaling systems.

We regard this criticism as mainly a formal one, which actually does little to undermine support for honest-signaling models. We concede that it would be better to have direct rather than indirect evidence of fitness costs. Nevertheless, we are comfortable with the assumption that indirect costs lead to fitness differences, at least for most of the kinds of costs that have been described. For example, given that the total energy available to any organism is limited, it seems logical to assume that allocation of some of that energy to signaling necessarily means that less is available for other functions that contribute to survival and reproduction. Thus it seems legitimate to assume that an energy cost will lead to a fitness cost, even though an empirical demonstration of such a connection is available in only a few instances (e.g., Mappes et al. 1996, Kotiaho 2000). Similarly, it seems reasonably safe to assume that fitness costs will result from decreased immune defenses, impairments of flight performance, increased vulnerability to attack, and so forth.

But Kotiaho (2001) raises another criticism of the evidence for signaling costs, one that is more troubling. Models that use costs to stabilize signals of quality or aggressiveness assume that marginal costs are greater for signalers of low quality than for signalers of high quality. As Kotiaho (2001) points out, very few studies have tested this assumption of "differential costs." Kotiaho's (2000) study of a wolf spider, *Hygrolycosa rubrofasciata*, provides one of the rare exceptions, and can be used to illustrate the difficulty of adequately demonstrating differential costs. Male spiders of this species drum their abdomen on dry leaves, producing sounds audible to humans. Because female wolf spiders prefer to mate with males that drum at higher rates (Kotiaho et al. 1996, 1998), the drumming can be assumed to be at least in part a sexual display. Males that are induced to drum at high rates have lower survival than males that are not induced to drum (Mappes et al. 1996), indicating that drumming has a fitness cost. To test whether this cost falls differentially on males in poor condition, Kotiaho (2000) simultaneously manipulated condition and drumming rate. To manipulate condition, males were given a high, medium, or low level of food. To manipulate drumming, males were exposed or not exposed to females. Exposure to females increased drumming rate approximately tenfold, and in all food-treatment groups the males exposed to females had lower survival. These results support the idea of a direct fitness cost of drumming, but note that it is possible that exposure to females induces other changes in male behavior or physiology besides drumming rate, and that these other changes, not drumming, are responsible for the observed effects on survival.

Accepting for the moment that drumming does have a survival cost, how do we use Kotiaho's (2000) experiment to test for differential costs? The crucial point is whether the change in survival due to exposure to females is greater among males in poor condition, that is, those given a low level of food, than among males in good condition, that is, those given a high level of food. Kotiaho (2000) does not test this point directly. Inspection of the data suggests that the results for large males support the prediction, but that the results on small males do not. Overall, then, support for differential costs in *H. rubrofasciata* is ambiguous.

Kotiaho (2001) cites Møller and de Lope (1994) as providing another test of differential costs, in this case for long tails in barn swallows. Møller and de Lope (1994) examined the survival of males following tail manipulation, combining data from four different experiments. Survival for each treatment group was expressed relative to the mean survival in a given year, to control for year-to-year variability. Overall, mean survival was highest for males whose tails were shortened, intermediate for control males, and lowest for males whose tails were elongated. These data again provide evidence for a direct fitness cost of long tails. The crucial point with respect to differential costs is that among males with elongated tails, those that survived had longer original tail lengths than males that did not survive. Among controls, differences in original tail length between survivors and nonsurvivors were much smaller. These results imply that tail elongation has a greater effect on survival in males with short tails than on survival in males with long tails. If we assume that original tail length reflects male quality, we can conclude that males of low quality pay a higher cost for an enhanced signal than do males of high quality.

The barn swallow study supports the assumption that males of different quality pay differential costs for having long tails, but for most signals we have no evidence that bears on the issue, and we are limited to evaluating the plausibility of the differential costs assumption. Certainly, differential costs make sense in some situations, such as when a signal reveals condition in the sense of energy balance and the signal's principal cost is energy expenditure. Here it seems inescapable that devoting a specified amount of energy to signaling will have less effect on the fitness of an individual in good condition than on the fitness of an individual in poor condition. The assumption is less plausible in other situations, as for example when the principal cost of a signal is an increase in predation risk. In this context, the differential cost assumption boils down to assuming that a given signal increases the risk of predation more in a low-quality signaler than in a high-quality one—for example that a given area of carotenoid pigmentation increases mortality due to predation more in a low-quality male than in a high-quality male. Although this assumption seems possible, it is not obviously true as a general proposition.

Overall, support for the kind of cost structure assumed by the handicap principle seems good for some signals and poor for others. The best case can be made for long tails in birds; here the signal has considerable costs, direct fitness costs have been shown, and there is some evidence that the costs fall differentially on low-quality signalers. At the other end of the spectrum are cases such as begging, for which costs in general seem unexpectedly low and no evidence exists that costs fall differentially on different categories of signalers.

Alternatives to the Handicap Mechanism

The handicap mechanism is not the only explanation available to account for signal reliability. One alternative allowed by current theory is that signals can be honest without cost if the interests of signaler and receivers do not conflict. Theory allows such an explanation, for example, if the signaler and receiver are genetic relatives and the information that is passed between them is limited in scope (Maynard Smith 1991a, Bergstrom and Lachmann 1998). This general category of explanation may apply to most cases of alarm signals. Existing evidence shows that alarm signals given by animals such as rodents, primates, and birds are somewhat reliable as concerns the proximity of a predator, and in some cases also contain reliable information on the category of predator or the degree of danger. The only appreciable cost of alarming is that the attention of the predator may be drawn to the alarmer. This cost should be negligible in cases where a predator has not in fact been observed, and thus the cost does not apply to false alarms. A cost that does not apply to deceptive signals cannot be used to explain signal honesty. Instead, reliable signaling is explained by the common interests of signaler and receiver. Often, signaler and receiver are genetic relatives, so the inclusive fitness of both increases if the signaler successfully warns the receiver of a predator's approach. Another possibility is that the signaler benefits, in terms of its own safety, when receivers scatter in response to the alarm, the receivers also benefiting from their responses (Charnov and Krebs 1975, Sherman 1985). In both these cases, signalers and receivers have common interests in the receivers being informed about a predator's approach, and the signal can be honest without being costly.

Food calls provide another case where signal reliability may sometimes be explained by the common interests of signalers and receivers. A signaler can benefit from attracting other individuals to a food source if they are its genetic relatives, if grouping provides safety from predation, or if a coalition is needed to protect the food. If one of these benefits applies, and assuming that the receivers benefit from approaching the food, then signalers and receivers again have common interests, and signals that are not costly can be reliable.

A second alternative to the handicap mechanism is a mechanism in which the benefits of signaling, rather than the costs, vary with signaler attributes (see figure 1.1b). Signals still have a cost, but it is the variation in signal benefits between signalers that produces signaling at different levels. This possibility, which is compatible with Grafen's mathematics, is central to many of the models of honest signaling of need discussed in chapter 2. In such models, the costs needed to produce equilibrium signaling solutions are in some cases projected to be rather formidable; nevertheless, it is the relationship between benefit and need that produces a reliable relationship between need and signaling level. The assumption that benefit varies with need seems entirely justified for cases where offspring are soliciting food; for example, it is logical to assume that the benefit of a given amount of food will be greater for an individual approaching starvation than for an individual that is already well fed.

Whether benefits can assume the relationships with signaler attributes necessary to produce honest signaling in other situations seems more doubtful, and has not been explored extensively. For mating signals, as an example, what would be needed is for the benefit of mating with a given female to be greater for a male of high quality than for a male of low quality, which seems questionable as a general proposition. Similarly, to produce honest signals of fighting ability, what is needed is for males of high fighting ability to benefit more from a contested resource than would males of low fighting ability; again, this is not an attractive assumption. Where variable benefits might be a more satisfying explanation of reliability is in signals of aggressiveness. A common explanation for why aggressiveness varies in the first place is the existence of a value asymmetry, meaning that one opponent values the contested resource more than does the other (Maynard Smith and Parker 1976). If differences in aggressiveness are based on value asymmetries, and value asymmetries are based on which opponent would benefit more from the resource, then it follows logically that the contestant of greater aggressiveness benefits more, and an honest signaling system can be built upon that relationship. Differences in aggressiveness can be explained, however, by other types of asymmetries that would not lead to the benefit of winning being higher for more aggressive individuals—for example asymmetries in fighting ability and arbitrary asymmetries (Maynard Smith and Parker 1976).

A third alternative is the "constraints" hypothesis, the idea that certain signals are forced to be reliable because of physiological or anatomical constraints on signal production. A signal often cited as illustrating such a constraint is dominant frequency in the calls of frogs and toads. The argument is that because dominant frequency is determined by size of the vocal apparatus (in this case by vocal-cord mass), and the size of the vocal apparatus is determined by body size, dominant frequency is constrained to contain reliable information about size. As we have seen (chapter 4), there are problems with this argument. For one, animals with a given vocal-cord mass can alter their dominant frequencies behaviorally, by changing pulmonary air pressure or vocal-cord tension. For another, animals of a given body size can alter their dominant frequencies developmentally, by producing vocal cords of differing masses. In other words, it is possible to manipulate both the relationship between the signal-producing mechanism and the signal property and the relationship between the signal-producing mechanism and the signaler attribute of interest. Given these opportunities for manipulation, it is not surprising that the correlations between body size and dominant frequency usually are not terribly strong in frogs and toads, with only about a third of the variance in dominant frequency explained by body size on average.

Because anurans have various options for altering the dominant frequency of their calls independently of body size, we cannot regard body size as constraining dominant frequency in any absolute sense. Instead, we think it preferable to view the reliability of dominant frequency as being maintained by the costs of exaggerating this signal attribute, as is the case in the handicap mechanism. Thus a male toad of a given size could produce calls of lower frequency than its peers by shunting more resources toward the development of especially massive vocal cords, but only at the cost of sacrificing the development of other aspects of its phenotype. Or a male could lower its dominant frequency by lowering the pressure difference between its lungs and its vocal cavity, but only at the cost of reducing the amplitude and thus the distance over which the call can be heard.

A more promising example of a constrained signal is that offered by formant dispersion, the spacing between the frequencies emphasized by vocal-tract resonances. Formant dispersion is determined in part by the length of the vocal tract, and the length of the vocal tract is in turn determined in part by body size (Fitch 1997, Fitch and Hauser 2003). These considerations generate an expectation of a negative correlation between formant dispersion and body size, and this expected correlation has been shown to exist in some mammals. Nevertheless, various mechanisms are known that could allow formant dispersion to be manipulated independently of body size. Behaviorally, a mammal can lengthen its vocal tract, and thus change formant dispersion, by pulling down on the larynx, by extending the head and neck, and by closing the mouth and pursing the lips. Developmentally, a mammal can lengthen its vocal tract by lowering its larynx in its throat, by producing an elongated proboscis, or by elongating the pharyngeal and oral cavities. Fitch and Hauser (2003) have argued that altering the morphology of the vocal tract is difficult in mammals, because of the various other functions performed by the anatomical elements of the vocal tract. Another way of saying this is that altering the vocal tract to manipulate formant dispersion would have considerable developmental costs.

Maynard Smith and Harper (2003) have argued that many signals are made reliable by constraints that make them "impossible to fake," but they acknowledge that in practice it may be difficult to differentiate such "indices of quality" from handicaps, which are honest because they are costly. Maynard Smith and Harper (2003) use formant dispersion in the roar of the red deer as a "paradigm example" of an index, whereas we prefer to view this signal as a handicap, made honest by a mix of developmental and production costs. The different labels applied to this signal (index versus handicap) should not obscure the fact that there is substantial agreement about the underlying reasons that the signal is honest. Agreement exists that: (1) the structure of the signal-production mechanism determines in part the properties of the signal, (2) the structure of the signal-production mechanism can be manipulated during development at some cost, (3) the signal properties of interest can be manipulated behaviorally as well as developmentally, again at some cost, and (4) receivers, when interpreting the signal, may be able to correct, in part at least, for behavioral manipulations. If agreement exists on the mechanisms that underlie the enforcement of signal reliability, not too much importance should be given to how the signal is labeled. Nevertheless, we suggest that it is problematic to present any signal as a paradigm example of signals that cannot be faked while acknowledging that mechanisms exist by which it can be and is faked.

The possible constraints on call frequency and formant dispersion are examples of internal constraints, in the sense that in each case it is the animal's own morphology that is constraining the signal properties. Signal properties may instead be limited in some cases by external constraints, that is, by constraints imposed by the external environment. A possible example concerns carotenoid pigmentation. We recognize that carotenoid coloration is in part limited by costs. Use of carotenoids for display might be costly, because carotenoids so used cannot then be employed in immune function; hence, carotenoid display advertises immunocompetence by "using up" some of the characteristic being advertised. In addition, carotenoid display is costly because of the energy needed to transport and process carotenoid pigments, so that carotenoid coloration would advertise energy reserves, again by using up some of the trait being advertised. In addition, however, carotenoid display is limited by the amount of carotenoid that can be obtained in the diet, that is, from the external environment. Again, one can think of this limitation in terms of a cost rather than a constraint: more carotenoid could be obtained by devoting more time to foraging, but only at the cost of devoting less time to other activities. Whether the cost or the constraint interpretation is preferred is largely a matter of taste.

A fourth alternative to the handicap principle is individually directed skepticism, as we discussed in chapter 2. In this mechanism, receivers remember the reliability of past signals directed at them by particular signalers and adjust their current response accordingly. Receivers thus are less likely to respond to signals from an individual that has deceived them in the past, and more likely to respond to an individual that has previously signaled reliably. This mechanism favors signal reliability, because the benefit a signaler might experience from deceiving a receiver in one interaction can be more than outweighed by the cost of lowering the receiver's probability of response in future interactions. Because individually directed skepticism can be stated as a cost, this mechanism can be considered as another variation on the handicap principle; however, the idea is so far from the original spirit of the handicap principle, in which signals produce costs because of their extravagance, that we suggest this should be regarded as a separate hypothesis.

The effectiveness of individually directed skepticism in enforcing reliability would be greatly increased if the mechanism could be extended from dyadic exchanges to eavesdropping, producing what we have termed "third-party skepticism." Again, the idea here is that individuals not directly involved in a signaling interaction would observe both the signals given and the interaction outcome, learn something about the reliability of one or both of the interactants, and use this information in shaping their own subsequent responses to those signalers. This idea, though attractive, is thus far purely speculative. Evidence exists that animals directly involved in signaling interactions can learn the reliability of the signals addressed to them, for example in the case of intergroup calling in vervet monkeys and food calling in domestic chickens. In addition, it is now well established that third parties eavesdrop on signaling interactions in certain systems, such as aggressive signaling in fish and singing interactions in birds. In the latter case, third-party observers of both sexes have been shown to make use of information contained within singing interactions, the males using the information to modulate their aggressive responses to the interactants, and the females to modulate their reproductive responses. Thus it is known that animals in some cases can learn the reliability of individuals that they have interacted with, and that in other cases animals can extract information from signal exchanges that they have observed as third parties, but it is not yet known whether these two processes can be combined, that is, whether animals ever learn the reliability of signalers by eavesdropping on their interactions.

In summary, the reliability of some classes of signals seems best explained by the handicap principle, in the sense of signal costs that act differentially on different categories of signalers. The best examples of signals that fall into this category are carotenoid coloration in fishes and birds and long tails in birds. Likely additional examples are structural features of bird song and fundamental frequency in frog calls, for both of which the relevant costs are probably developmental. For other classes of signals, reliability is best explained by the common interests of signaler and receiver. The best examples here are the alarm calls of rodents, primates, and birds, and the food calls of primates and birds. Differential benefits for different categories of signalers explain the reliability of other categories of signals, notably begging in the young of birds and mammals. Skepticism directed toward individual signalers, either by their intended receivers or by eavesdroppers, may also help explain the reliability of some signals, such as food calls in chickens and affiliative calls in primates. Clearly, the explanation of signal reliability requires multiple hypotheses rather than one. Although these various hypotheses are successful in explaining the reliability of many of the best-studied signals, the reliability of others remains a mystery.

Deceit

"Lord, lord, how this world is given to lying." (Falstaff in *Henry IV, Part I,* Act V, Scene iv)

Doubt has been expressed in the past about whether animal communication can ever be deceptive. In part, this doubt was motivated by skepticism concerning whether animals are capable of the cognitive feats necessary to meet a definition of intentional deception, one that requires a deceiver to intend to create a false belief in a receiver. In this book, we have used a functional definition of deception, which requires only that a signaler benefits from breaking the usual correlation between a characteristic of the signal and an attribute of the signaler or its environment, and which specifies nothing about cognitive processes. Because functional deception can be accomplished with little in the way of cognitive skills, objections to imputing cognitive abilities to other animals become moot. Another source of skepticism about deception was the once widespread assumption that communication within a species would primarily be cooperative; because of this expectation, interspecific deception was accepted much more readily than deception within a species. Over the past few decades, individual selection, rather than group selection, has been increasingly accepted as the primary engine of adaptation, making cooperation appear less likely, and intraspecific deception conversely seem more probable.

In the first generation of honest signaling models, such as Enquist's (1985) model of aggressive signaling, Grafen's (1990a,b) model of signals of mate quality, and Maynard Smith's (1991a) Sir Philip Sidney game, signals were predicted to be uniformly honest if they existed at all. The mathematics of these models might be simple or complex, but regardless, the models themselves were simple in the sense that they minimized both the number of dimensions on which signalers could differ from each other and the number of variables that affected the optimal choice of signaling level for a given signaler. Later models tended to be more complex in these senses, and with this added complexity, the possibility that deception could coexist with reliability sometimes emerged.

The Sir Philip Sidney game (chapter 2) well illustrates this progression. In the original SPS model (Maynard Smith 1991a), two classes of signalers were posited, needy and not needy. The signalers could solicit a resource from a related individual, and donation of that resource would raise the survival of the signaler and lower that of the donor. If the signal had a cost, signaling equilibria were possible even if a conflict of interest existed, such that the fitness of a nonneedy signaler was raised by transfer of the resource while the fitness of the donor was lowered. At the signaling equilibria, only needy individuals signaled, so the signal was entirely honest. Whereas in this first SPS model, signalers differed only in need, a second-generation model allowed signalers to differ also in their relatedness to the donor, the cost that they paid for signaling, or both (Johnstone and Grafen 1993). With these extra complications, signaling equilibria were possible in which the signal of need was sometimes given honestly and sometimes deceptively. Receiver response could be maintained at these equilibria as long as the proportion of dishonest signalers was not too high. It is the existence of classes of signalers differing in their relatedness to the donor, and in the costs they pay for signaling, that allows honest and deceitful signaling to coexist at equilibrium in the Johnstone and Grafen (1993) SPS model. Because of the differences in relatedness and costs, honesty can be optimal for one class of signaler while deceit is optimal for the other; because of the existence of honest signalers, receiver response to the signal can still be favored despite the existence of deceitful signalers.

Although Johnstone and Grafen's (1993) model illustrates important general principles of deceptive signaling, its assumptions are not met by any examples of deception that we can point to in actual signaling systems. Other second-generation models, however, do allow deceptive signaling under assumptions that fit real-world examples. In particular, Godfray's (1995) model of begging by two offspring introduces the complication that the condition of one offspring can influence the optimal begging level of the other. The model predicts that, holding the condition of the focal offspring constant, the begging intensity of that individual ought to increase as the condition of the second offspring deteriorates. This prediction has been supported empirically for American robins (Smith and Montgomerie 1991a) and yellow-headed blackbirds (Price et al. 1996), though results have been negative for European starlings (Kacelnik et al. 1995, Cotton et al. 1996). In zebra finches, the begging level of focal young has been shown to increase when they are exposed to playback of begging by other young (Muller and Smith 1978), which also supports the idea that offspring exaggerate their begging when in competition with others for parental resources. Exaggeration of signaling level when in competition may not seem a particularly satisfying case of deception, but we argue that it does meet the definition given for functional deception: a signaler breaks the usual correlation between the signal and some internal or external variable in a way that produces some benefit to itself.

One reason that competitive begging may seem somewhat unconvincing as an example of deception is that deception in this case requires only quantitative exaggeration of a signal that would be produced even if the signaler were being completely honest. Deception is more clear cut when a dishonest signal is given in a context in which an honest signal would not be given at all. Among signals that are given between individuals with overlapping interests, false alarms provide the best examples of such clear-cut deception. In great tits (Møller 1988a), white-winged shrike-tanagers, and bluish-slate antshrikes (Munn 1986), alarm calls are given in the absence of predators, and in circumstances in which the signaler benefits by moving its competitors away from sources of food. False alarms are also given by barn swallows (Møller 1989a) and Formosan squirrels (Tamura 1995), in these cases in circumstances in which the false alarms hamper sexual rivals of the signalers. For great tits and shrike tanagers, estimates have been made of the relative frequency of false and true alarms, and in both cases false alarms appear to be more common than true ones. Presumably, selection favors continued response to all alarms because receivers are unable to discriminate false from true alarms, and because the fitness consequences of failing to respond to a true alarm can be drastic.

Kokko (1997) developed models of mating signals in which some level of deception is possible, owing to the introduction of age structure as a complicating factor. The models assume that there is a trade-off between investment in survival and investment in advertisement, and that signalers are able to carry over some part of their investment in advertising from one age interval to the next; with these assumptions, it is possible for optimal signaling strategies to differ between age classes. Because signalers come in multiple age classes, equilibria are possible in which some age classes signal dishonestly about their quality, and yet receiver response is maintained because signalers in other age classes are honest. Kokko (1998) went on to demonstrate that classes of males in very poor condition might nevertheless advertise good condition, because the survival cost of advertisement matters less to those who are highly likely to die anyway.

Convincing empirical proof of dishonesty in mating signals is always likely to be difficult to come by because of our uncertainty about what constitutes mate quality in any given situation. Nevertheless, one good candidate case already exists for deception in signals of mate quality, that involving carotenoid pigmentation in three-spined sticklebacks. Candolin (1999) showed that the relationship between area of red and condition was U-shaped in a sample of male sticklebacks, such that the reddest males included those in the worst condition as well as those in the best condition. Males deprived of food increased their area of red, rather than decreased it. Candolin (2000) also showed that area of red correlated with the quality of parental care provided by males, and that the tightness of this correlation (and hence the honesty of the signal) increased when the males were held in groups rather than singly. These and other results support the interpretation that carotenoid coloration signals the condition of males and hence their quality as parents; that investment in carotenoid coloration is limited by survival costs that are mediated in part by the reaction of other males to high levels of signaling; and that certain males, in very poor condition, exaggerate their signal regardless of the cost because their chances of survival are so low anyway. This empirical case, then, corresponds in spirit if not in all of its details with Kokko's (1997, 1998) models of mixed honest and dishonest mating signals.

Models of aggressive signaling that allow mixtures of honesty and deceit are provided by Adams and Mesterton-Gibbons (1995) and by Számadó (2000). In the Adams and Mesterton-Gibbons (1995) model, the signalers are territory owners who differ in fighting strength. Owners can threaten challengers, who then decide whether or not to attack. Signaling equilibria are possible if the threat has a vulnerability cost, one that is imposed only if the opponent attacks and the owner loses the resulting fight. Fighting costs are assumed to increase as fighting strength decreases. Adams and Mesterton-Gibbons (1995) find that in all equilibrium solutions threats are given by owners of very low fighting strength as well as by owners of very high fighting strength. The reason that very weak owners threaten despite their high average signal costs is that they also benefit more from threats than do other individuals. Benefits are high for these weak individuals because bluffing is the only way that they can win the resource, and because avoiding fights is especially advantageous to them, owing to the high costs they must pay if they fight. At equilibrium, some receivers continue to respond to threats, despite the occurrence of dishonest threats, because of the high cost of failing to heed the honest threats given by the strongest individuals.

The Adams and Mesterton-Gibbons (1995) model was formulated to match the best-established real-world case of deceit in aggressive contexts—the use of meral-spread displays by newly molted stomatopods of the species *Gonodactylus bredini* (Steger and Caldwell 1983, Adams and Caldwell 1990). Meral spreads show off the largest segments of the raptorial appendages, which are used by stomatopods in fighting, and seem to signal both intention to attack and fighting ability. A meral spread can be regarded as an honest threat when given by most cavity owners, because the signaler is capable of dealing a damaging blow and is likely to attack if the receiver fails to back down. The same display is deceptive when given by newly molted individuals, however, since these are unable to strike an opponent, owing to the weakness of their exoskeletons. Meral spreads are given more frequently per encounter by newly molted individuals than by intermolts, who are more likely to proceed directly to attack without bothering to threaten (Steger and Caldwell 1983). Nevertheless, because most individuals in a population are between molts at any given time, over 80% of meral spreads are given by individuals that are capable of striking (Steger and Caldwell 1983). Receiver response to the signal presumably is maintained by the high proportion of honest displays and by the potentially high cost of ignoring an honest threat. Deception is, again, particularly clear cut because newly molted individuals, if honest, would not signal at all.

Aggressive signaling also provides some more subtle examples of deception, wherein animals merely exaggerate their strength or aggressiveness. One such case occurs in the snapping shrimp studied by Hughes (1996, 2000). Here, body size is crucial in deciding aggressive contests, and the size of the major chela is used to signal body size. Individuals that have major chelae larger than expected for their body size take advantage of this fact by displaying their chelae more often than do other individuals. Another possible case of deception through exaggeration involves the dominant frequency of calls in Blanchard's cricket frogs (Wagner 1989,1992, Burmeister et al. 2002). Dominant frequency is inversely proportional to body size, which is again of great importance to the outcome of aggressive contests. Males lower their dominant frequencies when entering aggressive encounters, thus feigning larger size.

Deception tends to emerge in signaling models as they become more complex in terms of the number of dimensions on which signalers are allowed to differ, and in the number of variables that are allowed to affect the optimal choice of signaling level. Complexity in these senses is entirely realistic, in the sense that real-world biological signaling systems are bound to be more complex than the most complex signaling models. Thus we believe that it is realistic to expect deception to be widespread in animal signaling systems. Evidence supporting the occurrence of deception has been found in all the major categories of signaling systems that we have discussed, including begging, alarming, mating signals, and aggressive signals. Admittedly, well-demonstrated cases of deception are rare in relation to the number of signaling systems that have been described, but we believe that the rarity of good cases is due largely to the difficulties of adequately demonstrating deception when it occurs. These difficulties are particularly acute when deception takes the form of exaggeration or bluff, which is how deception is likely to manifest itself whenever signals are graded rather than dichotomous. Further work on exaggeration and bluff in animal communication is much needed.

The Balance of Reliability and Deceit

The signaling systems we have reviewed have proven to be largely reliable, with some admixture of deceit. This ought not to be surprising. Signaling systems will not be stable unless they are in large part reliable, or "honest on average," so that receivers will continue to respond to the signals, and signaling will continue to be favored. Given the complexities of real-life signaling systems, we can expect that in many cases some fraction of signalers will be able to take advantage of the systems by signaling dishonestly. Deceptive signalers will often be in a minority, but that need not necessarily be the case; deception can be the majority strategy if the cost to receivers of failing to respond to honest signals is sufficiently high.

Deceit in animal communication depends on reliability in a couple of ways. The first is methodological, and thus somewhat trivial. In general, we recognize the honest meaning of a signal by measuring a correlation between the signal and some attribute of the signaler or its environment. Only if we have established this correlation can we identify cases where the correlation is broken in ways that benefit the signaler. The correlation constitutes reliability and its breakdown constitutes deceit; thus we are able to identify deceit only if we have first established reliability.

More fundamentally, deceit depends on reliability, because reliability is necessary if receivers are ever to respond to a signal at all. This is a point that we have come back to over and over again throughout the course of this book. Only if the signal is reliable often enough to make response beneficial on average will receivers continue to respond. Only if receivers respond can the production of a deceptive signal be beneficial to a signaler. These precepts mesh not only with theory but with observation as well; wherever deception is found in animal communication, it coexists with and depends on reliability.

References

Adams, E. S., and R. L. Caldwell. 1990. Deceptive communication in asymmetric fights of the stomatopod crustacean *Gonodactylus bredini. Anim. Behav.* 39:706–716.

Adams, E. S., and M. Mesterton-Gibbons. 1995. The cost of threat displays and the stability of deceptive communication. *J. Theor. Biol.* 175:405–421.

Airey, D. C., K. L. Buchanan, T. Székely, C. K. Catchpole, and T. J. DeVoogd. 2000. Song, sexual selection, and a song control nucleus (HVC) in the brains of European sedge warblers. *J. Neurobiol.* 44:1–6.

Airey, D. C., and T. J. DeVoogd. 2000. Greater song complexity is associated with augmented song system anatomy in zebra finches. *Neuroreport* 11:2339–2344.

Alatalo, R. V., C. Glynn, and A. Lundberg. 1990. Singing rate and female attraction in the pied flycatcher: an experiment. *Anim. Behav.* 39:601–603.

Andersson, M. 1976. Social behaviour and communication in the great skua. *Behaviour* 58:40–77.

Andersson, M. 1982. Female choice selects for extreme tail length in a widowbird. *Nature* 299:818–820.

Andersson, M. 1986. Evolution of condition-dependent sex ornaments and mating preferences: sexual selection based on viability differences. *Evolution* 40:804–816.

Andersson, M. 1994. *Sexual Selection.* Princeton Univ. Press, Princeton, NJ.

Andersson, S. 1992. Female preferences for long tails in lekking Jackson's widowbirds: experimental evidence. *Anim. Behav.* 43:379–388.

Aparicio, J. M., R. Bonal, and P. J. Cordero. 2003. Evolution of the structure of tail feathers: implications for the theory of sexual selection. *Evolution* 57:397–405.

Arak, A. 1983. Sexual selection by male-male competition in natterjack toad choruses. *Nature* 306:261–262.

Bachman, G. C., and M. A. Chappell. 1998. The energetic cost of begging behaviour in nestling house wrens. *Anim. Behav.* 55:1607–1618.

Backwell, P. R. Y., J. H. Christy, S. R. Telford, M. D. Jennions, and N. I. Passmore. 2000. Dishonest signalling in a fiddler crab. *Proc. R. Soc. Lond. B* 267:719–724.

Bailey, W. J. 1991. *Acoustic Behaviour of Insects: An Evolutionary Perspective.* Chapman and Hall, London.

Baker, M. C. 1983. The behavioral response of female Nuttall's white-crowned sparrows to male song of natal and alien dialects. *Behav. Ecol. Sociobiol.* 12:309–315.

Baker, M. C., A. E. M. Baker, M. A. Cunningham, D. B. Thompson, and D. F. Tomback. 1984. Reply to "allozymes and song dialects: a reassessment." *Evolution* 38:449–451.

Baker, M. C., T. K. Bjerke, H. Lampe, and Y. Espmark. 1986. Sexual response of female great tits to variation in size of males' song repertoires. *Am. Nat.* 128:491–498.

Baker, M. C., and M. A. Cunningham. 1985. The biology of bird-song dialects. *Behav. Brain Sci.* 8:85–133.

Baker, M. C., K. J. Spitler-Nabors, and D. C. Bradley. 1981. Early experience determines song dialect responsiveness of female sparrows. *Science* 214:819–821.

Baker, M. C., K. J. Spitler-Nabors, A. D. Thompson, Jr., and M. A. Cunningham. 1987. Reproductive behaviour of female white-crowned sparrows: effect of dialects and synthetic hybrid songs. *Anim. Behav.* 35:1766–1774.

Baker, M. C., D. B. Thompson, G. L. Sherman, M. A. Cunningham, and D. F. Tomback. 1982. Allozyme frequencies in a linear series of song dialect populations. *Evolution* 36:1020–1029.

Bakker, T. C. M., and B. Mundwiler. 1994. Female mate choice and male red coloration in a natural three-spined stickleback (*Gasterosteus aculeatus*) population. *Behav. Ecol.* 5:74–80.

Ballentine, B., J. Hyman, and S. Nowicki. 2004. Vocal performance influences female response to male bird song: an experimental test. *Behav. Ecol.* 15:163–168.

Balmford, A., A. L. R. Thomas, and I. L. Jones. 1993. Aerodynamics and the evolution of long tails in birds. *Nature* 361:628–631.

Balph, D. M., and D. F. Balph. 1966. Sound communication of Uinta ground squirrels. *J. Mamm.* 47:440–450.

Balph, M. H., D. F. Balph, and H. C. Romesburg. 1979. Social status signaling in winter flocking birds: an examination of a current hypothesis. *Auk* 96:78–93.

Balsby, T. J. S., and T. Dabelsteen. 2002. Female behaviour affects male courtship in whitethroats, *Sylvia communis*: an interactive experiment using visual and acoustic cues. *Anim. Behav.* 63:251–257.

Baptista, L. F., and L. Petrinovich. 1986. Song development in the white-crowned sparrow: social factors and sex differences. *Anim. Behav.* 34:1359–1371.

Barrowclough, G. F. 1980. Gene flow, effective population sizes, and genetic variance components in birds. *Evolution* 34:789–798.

Basolo, A. L. 1990. Female preference predates the evolution of the sword in swordtail fish. *Science* 250:808–810.

Basolo, A. L. 1995. Phylogenetic evidence for the role of a pre-existing bias in sexual selection. *Proc. R. Soc. Lond. B* 259:307–311.

Bateman, A. J. 1948. Intra-sexual selection in *Drosophila. Heredity* 2:349–368.

Batteau, D. W. 1968. The world as a source; the world as a sink. Pp. 197–203 in S. J. Freedman, ed., *The Neuropsychology of Spatially Oriented Behavior*. Dorsey Press, Homewood, IL.

Bee, M. A. 2002. Territorial male bullfrogs (*Rana catesbeiana*) do not assess fighting ability based on size-related variation in acoustic signals. *Behav. Ecol.* 13:109–124.

Bee, M. A., and H. C. Gerhardt. 2001. Neighbour-stranger discrimination by territorial male bullfrogs (*Rana catesbeiana*): II. Perceptual basis. *Anim. Behav.* 62:1141–1150.

Bee, M. A., and S. A. Perrill. 1996. Responses to conspecific advertisement calls in the green frog (*Rana clamitans*) and their role in male-male communication. *Behaviour* 133:283–301.

Bee, M. A., S. A. Perrill, and P. C. Owen. 1999. Size assessment in simulated territorial encounters between male green frogs (*Rana clamitans*). *Behav. Ecol. Sociobiol.* 45:177–184.

Bee, M. A., S. A. Perrill, and P. C. Owen. 2000. Male green frogs lower the pitch of acoustic signals in defense of territories: a possible dishonest signal of size? *Behav. Ecol.* 11:169–177.

Bell, G. 1978. The handicap principle in sexual selection. *Evolution* 32:872–885.

Bengtsson, H., and O. Rydén. 1983. Parental feeding rate in relation to begging behavior in asynchronously hatched broods of the great tit *Parus major*. *Behav. Ecol. Sociobiol.* 12:243–251.

Bennett, A. T. D., I. C. Cuthill, and K. J. Norris. 1994. Sexual selection and the mismeasure of color. *Am. Nat.* 144:848–860.

Bergman, T. J., J. C. Beehner, D. L. Cheney, and R. M. Seyfarth. 2003. Hierarchical classification by rank and kinship in baboons. *Science* 302:1234–1236.

Bergstrom, C. T., and M. Lachmann. 1997. Signalling among relatives. I. Is costly signalling too costly? *Phil. Trans. Roy. Soc. B* 352:609–617.

Bergstrom, C. T., and M. Lachmann. 1998. Signaling among relatives. III. Talk is cheap. *Proc. Natl. Acad. Sci. USA* 95:5100–5105.

Bernard, D. J., M. Eens, and G. F. Ball. 1996. Age- and behavior-related variation in volumes of song control nuclei in male European starlings. *J. Neurobiol.* 30:329–339.

Bildstein, K. L. 1983. Why white-tailed deer flag their tails. *Am. Nat.* 121:709–715.

Blount, J. D., N. B. Metcalfe, T. R. Birkhead, and P. F. Surai. 2003. Carotenoid modulation of immune function and sexual attractiveness in zebra finches. *Science* 300:125–127.

Blumstein, D. T., and K. B. Armitage. 1997. Alarm calling in yellow-bellied marmots: I. The meaning of situationally variable alarm calls. *Anim. Behav.* 53:143–171.

Blumstein, D. T., and K. B. Armitage. 1998. Why do yellow-bellied marmots call? *Anim Behav.* 56:1053–1055.

Blumstein, D. T., J. Steinmetz, K. B. Armitage, and J. C. Daniel. 1997. Alarm calling in yellow-bellied mamots. II. The importance of direct fitness. *Anim. Behav.* 53:173–184.

Boag, P. T. 1987. Effects of nestling diet on growth and adult size of zebra finches (*Poephila guttata*). *Auk* 104:155–166.

Boinski, S., and A. F. Campbell. 1996. The huh vocalization of white-faced capuchins: a spacing call disguised as a food call? *Ethology* 102:826–840.

Borror, J. 1965. Song variation in Maine song sparrows. *Wilson Bull.* 77:5–37.

Bottjer, S. W., S. L. Glaessner, and A. P. Arnold. 1985. Ontogeny of brain nuclei controlling song learning and behavior in zebra finches. *J. Neurosci.* 5:1556–1562.

Bottjer, S. W., E. A. Miesner, and A. P. Arnold. 1984. Forebrain lesions disrupt development but not maintenance of song in passerine birds. *Science* 224:901–903.

Bradbury, J. W., and S. L. Vehrencamp. 1998. *Principles of Animal Communication.* Sinauer, Sunderland, MA.

Brawner, W. R., G. E. Hill, and C. A. Sunderman. 2000. Effects of coccidial and mycoplasmal infections on carotenoid-based plumage pigmentation in male house finches. *Auk* 117:952–963.

Brenowitz, E. A., and D. E. Kroodsma. 1996. The neuroethology of birdsong. Pp. 285–305 in D. E. Kroodsma and E. H. Miller, eds., *Ecology and Evolution of Acoustic Communication in Birds.* Cornell Univ. Press, Ithaca, N.Y.

Brenowitz, E. A., B. Nalls, J. C. Wingfield, and D. E. Kroodsma. 1991. Seasonal changes in avian song control nuclei without changes in song repertoire. *J. Neurosci.* 11:1367–1374.

Brilot, B. O., and R. A. Johnstone. 2003. The limits to cost-free signalling of need between relatives. *Proc. R. Soc. Lond. B* 270:1055–1060.

Bronstein, P. M. 1984. Agonistic and reproductive interactions in *Betta splendens*. *J. Comp. Psych.* 98:421–431.

Brown, J. L. 1975. *The Evolution of Behavior.* Norton, New York.

Brown, W. D., J. Wideman, M. C. B. Andrade, A. C. Mason, and D. T. Gwynne. 1996. Female choice for an indicator of male size in the song of the black-horned tree cricket *Oecanthus nigricornis* (Orthoptera: Gryllidae: Oecanthinae). *Evolution* 50:2400–2411.

Brush, A. H., and D. M. Power. 1976. House finch pigmentation: carotenoid metabolism and the effect of diet. *Auk* 93:725–739.

Buchanan, K. L., and C. K. Catchpole. 1997. Female choice in the sedge warbler, *Acrocephalus schoenobaenus*: multiple cues from song and territory quality. *Proc. Roy. Soc. Lond. B* 264:521–526.

Buchanan, K. L., and C. K. Catchpole. 2000. Song as an indicator of male parental effort in the sedge warbler. *Proc. R. Soc. Lond. B* 267:321–326.

Buchanan, K. L., C. K. Catchpole, J. W. Lewis, and A. Lodge. 1999. Song as an indicator of parasitism in the sedge warbler. *Anim. Behav.* 57:307–314.

Buchanan, K. L., and M. R. Evans. 2000. The effect of tail streamer length on aerodynamic performance in the barn swallow. *Behav. Ecol.* 11:228–238.

Buchanan, K. L., and M. R. Evans. 2001. Flight, fitness, and sexual selection: a response. *Behav. Ecol.* 12:513–515.

Buchanan, K. L., M. R. Evans, and A. R. Goldsmith. 2003a. Testosterone, dominance signalling and immunosuppression in the house sparrow, *Passer domesticus. Behav. Ecol. Sociobiol.* 55:50–59.

Buchanan, K. L., M. R. Evans, A. R. Goldsmith, D. M. Bryant, and L. V. Rowe. 2001. Testosterone influences basal metabolic rate in male house sparrows: a new cost of dominance signalling? *Proc. R. Soc. Lond. B* 268:1337–1344.

Buchanan, K. L., K. A. Spencer, A. R. Goldsmith, and C. K. Catchpole. 2003b. Song as an honest signal of past developmental stress in the European starling (*Sturnus vulgaris*). *Proc. R. Soc. Lond. B* 270:1149–1156.

Burford, J. E., T. J. Friedrich, and K. Yasukawa. 1998. Response to playback of nestling begging in the red-winged blackbird, *Agelaius phoeniceus. Anim. Behav.* 56:555–561.

Burghardt, G. M. 1988. Anecdotes and critical anthropomorphism. *Behav. Brain Sci.* 11:248–249.

Burmeister, S. S., A. G. Ophir, M. J. Ryan, and W. Wilczynski. 2002. Information transfer during cricket frog contests. *Anim. Behav.* 64:715–725.

Byrne, R. W., and A. Whiten. 1988. Toward the next generation in data quality: A new survey of primate tactical deception. *Behav. Brain Sci.* 11:267–271.

Byrne, R. W., and A. Whiten. 1992. Cognitive evolution in primates: evidence from tactical deception. *Man* 27:609–627.

Caine, N. G., R. L. Addington, and T. L. Windfelder. 1995. Factors affecting the rates of food calls given by red-bellied tamarins. *Anim. Behav.* 50:53–60.

Caldwell, R. L. 1984. A test of individual recognition in the stomatopod *Gonodactylus festae. Anim. Behav.* 33:101–106.

Caldwell, R. L. 1986. The deceptive use of reputation by stomatopods. Pp. 129–145 in R. W. Mitchell and N. S. Thomson, eds., *Deception: Perspectives on Human and Nonhuman Deceit.* SUNY Press, Albany.

Caldwell, R. L., and H. Dingle. 1975. Ecology and evolution of agonistic behavior in stomatopods. *Naturwissenschaften* 62:214–222.

Canady, R. A., D. E. Kroodsma, and F. Nottebohm. 1984. Population differences in complexity of a learned skill are correlated with the brain space involved. *Proc. Natl. Acad. Sci. USA* 81:6232–6234.

Candolin, U. 1999. The relationship between signal quality and physical condition: is sexual signalling honest in the three-spined stickleback? *Anim. Behav.* 58:1261–1267.

Candolin, U. 2000. Male-male competition ensures honest signaling of male parental ability in the three-spined stickleback (*Gaterosteus aculeatus*). *Behav. Ecol. Sociobiol.* 49:57–61.

Caro, T. M. 1986a. The functions of stotting: a review of the hypotheses. *Anim. Behav.* 34:649–662.

Caro, T. M. 1986b. The function of stotting in Thomson's gazelles: some tests of the predictions. *Anim. Behav.* 34:663–684.

Caro, T. M. 1995. Pursuit-deterrence revisited. *Trends Ecol. Evol.* 10:500–503.

Caro, T. M., L. Lombardo, A. W. Goldizen, and M. Kelly. 1995. Tail-flagging and other antipredator signals in white-tailed deer: new data and synthesis. *Behav. Ecol.* 6:442–450.

Caryl, P. G. 1979. Communication by agonistic displays: what can games theory contribute to ethology? *Behaviour* 68:136–169.

Castro, C. A., and J. W. Rudy. 1987. Early-life malnutrition selectively retards the development of distal-cue but not proximal-cue navigation. *Devel. Psychobiol.* 20:521–537.

Catchpole, C. K. 1980. Sexual selection and the evolution of complex songs among European warblers of the genus *Acrocephalus*. *Behaviour* 74:149–166.

Catchpole, C. K., J. Dittami, and B. Leisler. 1984. Differential responses to male song in female songbirds implanted with oestradiol. *Nature* 312:563–564.

Catchpole, C. K., B. Leisler, and J. Dittami. 1986. Sexual differences in the responses of captive great reed warblers (*Acrocephalus arundinaceus*) to variation in song structure and repertoire size. *Ethology* 73:69–77.

Catchpole, C. K., and P. J. B. Slater. 1995. *Bird Song:Biological Themes and Variations*. Cambridge Univ. Press, Cambridge.

Chapman, C. A., and L. Lefebvre. 1990. Manipulating foraging group size: spider monkey food calls at fruiting trees. *Anim. Behav.* 39:891–896.

Chappell, M. A., and G. C. Bachman. 1998. Exercise capacity of house wren nestlings: begging chicks are not working as hard as they can. *Auk* 115:863–870.

Charnov, E. L., and J. R. Krebs. 1975. The evolution of alarm calls: altruism or manipulation? *Am. Nat.* 109:107–112.

Chase, I. D., C. Bartolomeo, and L. A. Dugatkin. 1994. Aggressive interactions and inter-contest interval: how long do winners keep winning? *Anim. Behav.* 48:393–400.

Cheney, D. L., and R. M. Seyfarth. 1988. Assessment of meaning and the detection of unreliable signals by vervet monkeys. *Anim. Behav.* 36:477–486.

Cheney, D. L., and R. M. Seyfarth. 1990. *How Monkeys See the World: Inside the Mind of Another Species*. Univ. Chicago Press, Chicago.

Cheney, D. L., R. M. Seyfarth, and J. B. Silk. 1995. The responses of female baboons (*Papio cynocephalus ursinus*) to anomalous social interactions: evidence for causal reasoning? *J. Comp. Psych.* 109:134–141.

Clark, A. B., and W.-H. Lee. 1998. Red-winged blackbird females fail to increase feeding in response to begging call playbacks. *Anim. Behav.* 56:563–570.

Clark, C. W., P. Marler, and K. Beeman. 1987. Quantitative analysis of animal vocal phonology: an application to swamp sparrow song. *Ethology* 76:101–115.

Clotfelter, E. D., K. A. Schubert, V. Nolan, and E. D. Ketterson. 2003. Mouth color signals thermal state of nestling dark-eyed juncos (*Junco hyemalis*). *Ethology* 109:171–182.

Clutton-Brock, T. H. 1982. The functions of antlers. *Behaviour* 79:108–125.

Clutton-Brock, T. H., and S. D. Albon. 1979. The roaring of red deer and the evolution of honest advertisement. *Behaviour* 69:145–170.

Clutton-Brock, T. H., and A. C. J. Vincent. 1991. Sexual selection and the potential reproductive rates of males and females. *Nature* 351:58–60.

Collias, N. E., and E. C. Collias. 1967. A field study of the red jungle fowl in north-central India. *Condor* 69:360–386.

Collins, S. A. 2000. Men's voices and women's choices. *Anim. Behav.* 60:773–780.

Collins, S. A., C. Hubbard, and A. M. Houtman. 1994. Female mate choice in the zebra finch—the effect of male beak colour and male song. *Behav. Ecol. Sociobiol.* 35:21–25.

Cotton, P. A., A. Kacelnik, and J. Wright. 1996. Chick begging as a signal: are nestlings honest? *Behav. Ecol.* 2:178–182.

Coyne, J. A., and H. A. Orr. 1997. "Patterns of speciation in *Drosophila*" revisited. *Evolution* 51:295–303.

Cunningham, M. A., and M. C. Baker. 1983. Vocal learning in white-crowned sparrows: sensitive phase and song dialects. *Behav. Ecol. Sociobiol.* 13:259–269.

Cuthill, I. C., J. C. Partridge, and A. T. D. Bennett. 2000. Avian UV vision and sexual selection. Pp. 61–82 in Y. Espmark, T. Amundsen, and G. Rosenqvist, eds., *Animal Signals: Signalling and Signal Design in Animal Communication.* Tapir Academic Press, Trondheim, Norway.

Dabelsteen, T., P. K. McGregor, J. Holland, J. A. Tobias, and S. B. Pedersen. 1997. The signal value of overlapping singing in male robins. *Anim. Behav.* 53:249–256.

Dabelsteen, T., P. K. McGregor, M. Shepherd, X. Whittaker, and S. B. Pedersen. 1996. Is the signal value of overlapping different from that of alternating during matched singing in Great Tits? *J. Avian Biol.* 27:189–194.

Darwin, C. 1859. *On the Origin of Species.* John Murray, London.

Darwin, C. 1871. *The Descent of Man, and Selection in Relation to Sex.* John Murray, London.

Davies, N. B., and T. R. Halliday. 1978. Deep croaks and fighting assessment in toads *Bufo bufo. Nature* 274:683–685.

Davies, N. B., R. M. Kilner, and D. G. Noble. 1998. Nestling cuckoos, *Cuculus canorus*, exploit hosts with begging calls that mimic a brood. *Proc. Roy. Soc. Lond. B* 265:673–678.

Davies, N. B., and A. Lundberg. 1984. Food distribution and a variable mating system in the dunnock, *Prunella modularis. J. Anim. Ecol.* 53:895–912.

Dawkins, M. S., and T. Guilford. 1991. The corruption of honest signalling. *Anim. Behav.* 41:865–873.

Dawkins, R. 1976. *The Selfish Gene.* Oxford Univ. Press, Oxford.

Dawkins, R., and J. R. Krebs. 1978. Animal signals: information or manipulation? Pages 282–309 in J. R. Krebs and N. B. Davies, eds., *Behavioural Ecology*. Blackwell, Oxford.

Day, T. 2000. Sexual selection and the evolution of costly female preferences: spatial effects. *Evolution* 54:715–730.

Dearborn, D. C. 1999. Brown-headed cowbird nestling vocalizations and risk of nest predation. *Auk* 116:448–457.

Dennett, D. C. 1988. Precis of The Intentional Stance. *Behav. Brain Sci.* 11:495–546.

DeVoogd, T. J., J. R. Krebs, S. D. Healy, and A. Purvis. 1993. Relations between song repertoire size and the volume of brain nuclei related to song: comparative evolutionary analyses amongst oscine birds. *Proc. R. Soc. Lond. B* 254:75–82.

Dingle, H. 1969. A statistical and information analysis of aggressive communication in the mantis shrimp *Gonodactylus bredini* Manning. *Anim. Behav.* 17:561–575.

Dingle, H. 1972. Aggressive behavior in stomatopods and the use of information theory in the analysis of animal communication. Pp. 126–155 in H. E. Winn and B. L. Olla, eds., *Behavior of Marine Animals: Current Perspectives in Research. I. Invertebrates*. Plenum, New York.

Dingle, H., and R. L. Caldwell. 1969. The aggressive and territorial behaviour of the mantis shrimp *Gonodactylus bredini* Manning (Crustacea: Stomatopoda). *Behaviour* 33:115–136.

Doutrelant, C., and P. K. McGregor. 2000. Eavesdropping and mate choice in female fighting fish. *Behaviour* 137:1655–1669.

Doutrelant, C., P. K. McGregor, and R. F. Oliveira. 2001. The effect of an audience on intrasexual communication in male Siamese fighting fish, *Betta splendens*. *Behavioral Ecology* 12:283–286.

Duckworth, R. A., M. T. Mendonca, and G. E. Hill. 2001. A condition dependent link between testosterone and disease resistance in the house finch. *Proc. R. Soc. Lond. B* 268:2467–2472.

Duffy, D. L., G. E. Bentley, D. L. Drazen, and G. F. Ball. 2000. Effects of testosterone on cell-mediated and humoral immunity in non-breeding adult European starlings. *Behav. Ecol.* 11:654–662.

Dunford, C. 1977. Kin selection for ground squirrel alarm calls. *Am. Nat.* 111:782–785.

Dunham, D. W. 1966. Agonistic behavior in captive rose-breasted grosbeaks, *Pheucticus ludovicianus* (L.). *Behaviour* 27:160–173.

Earley, R. L., and L. A. Dugatkin. 2002. Eavesdropping on visual cues in green swordtail (*Xiphophorus helleri*) fights: a case for networking. *Proc. R. Soc. Lond. B* 269:943–952.

East, M. 1981. Alarm calling and parental investment in the robin *Erithacus rubecula*. *Ibis* 123:223–230.

Eastzer, D. H., A. P. King, and M. J. West. 1985. Patterns of courtship between cowbird subspecies: evidence for positive assortment. *Anim. Behav.* 33:30–39.

Eberhardt, L. S. 1994. Oxygen consumption during singing by male Carolina wrens (*Thryothorus ludovicianus*). *Auk* 111:124–130.

Eens, M., R. Pinxten, and R. F. Verheyen. 1991. Male song as a cue for mate choice in the European starling. *Behaviour* 116:210–238.

Ekman, J. B., and C. E. H. Askenmo. 1984. Social rank and habitat use in willow tit groups. *Anim. Behav.* 32:508–514.

Elgar, M. A. 1986. House sparrows establish foraging flocks by giving chirrup calls if the resources are divisible. *Anim. Behav.* 34:169–174.

Elowson, A. M., P. L. Tannenbaum, and C. T. Snowdon. 1991. Food-associated calls correlate with food preferences in cotton-top tamarins. *Anim. Behav.* 42:931–937.

Endler, J. A. 1980. Natural selection on color patterns in *Poecilia reticulata*. *Evolution* 34:76–91.

Endler, J. A. 1983. Natural and sexual selection on color patterns in poeciliid fishes. *Environmental Biology of Fishes* 9:173–190.

Endler, J. A. 1990. On the measurement and classification of colour in studies of animal colour patterns. *Biol. J. Linn. Soc.* 41:315–352.

Endler, J. A., and A. L. Basolo. 1998. Sensory ecology, receiver biases and sexual selection. *Trends Ecol. Evol.* 13:415–420.

Enquist, M. 1985. Communication during aggressive interactions with particular reference to variation in choice of behaviour. *Anim. Behav.* 33:1152–1161.

Evans, C. S. 1985. Display vigor and subsequent fight performance in the Siamese fighting fish, *Betta splendens*. *Behav. Process.* 11:113–122.

Evans, C. S., and L. Evans. 1999. Chicken food calls are functionally referential. *Anim. Behav.* 58:307–319.

Evans, C. S., and L. Evans. 2002. Sceptical hens: receivers constrain deceptive food calling by male fowl. Animal Behavior Society Meeting, Indiana Univ., Bloomington, IN.

Evans, C. S., and L. Evans. 2003. Adaptive scepticism. Animal Behavior Society Meeting, Boise State Univ., Boise, ID.

Evans, C. S., L. Evans, and P. Marler. 1993a. On the meaning of alarm calls: functional reference in an avian vocal system. *Anim. Behav.* 46:23–38.

Evans, C. S., J. M. Macedonia, and P. Marler. 1993b. Effects of apparent size and speed on the response of chickens, *Gallus gallus*, to computer-generated simulations of aerial predators. *Anim. Behav.* 46:1–11.

Evans, C. S., and P. Marler. 1994. Food calling and audience effects in male chickens, *Gallus gallus*: their relationships to food availability, courtship and social facilitation. *Anim. Behav.* 47:1159–1170.

Evans, M. R. 1998. Selection on swallow tail streamers. *Nature* 394:233–234.

Evans, M. R., A. R. Goldsmith, and S. R. A. Norris. 2000. The effects of testosterone on antibody production and plumage coloration in male house sparrows (*Passer domesticus*). *Behav. Ecol. Sociobiol.* 47:156–163.

Evans, M. R., and B. J. Hatchwell. 1992. An experimental study of male adornment in the scarlet-tufted malachite sunbird: II. The role of the elongated tail in mate choice and experimental evidence for a handicap. *Behav. Ecol. Sociobiol.* 29:421–427.

Evans, M. R., and A. L. R. Thomas. 1992. The aerodynamic and mechanical effects of elongated tails in the scarlet-tufted malachite sunbird: measuring the cost of a handicap. *Anim. Behav.* 43:337–347.

Evans, M. R., and A. L. R. Thomas. 1997. Testing the functional significance of tail streamers. *Proc. R. Soc. Lond. B* 264:211–217.

Falls, J. B. 1982. Individual recognition by sounds in birds. Pp. 237–278 in D. E. Kroodsma and E. H. Miller, eds., *Acoustic Communication in Birds*. Academic Press, New York.

Feener, D. H., L. F. Jacobs, and J. O. Schmidt. 1996. Specialized parasitoid attracted to a pheromone of ants. *Anim. Behav.* 51:61–66.

Fisher, R. A. 1930. *The Genetical Theory of Natural Selection*. Clarendon Press, Oxford.

Fitch, W. T. 1997. Vocal tract length and formant frequency dispersion correlate with body size in rhesus macaques. *Journal of the Acoustic Society of America* 102:1213–1222.

Fitch, W. T., and M. D. Hauser. 2003. Unpacking "honesty": vertebrate vocal production and the evolution of acoustic signals. Pp. 65–137 in A. M. Simmons, R. R. Fay, and A. N. Popper, eds., *Acoustic Communication*. Springer, New York.

Fitch, W. T., and D. Reby. 2001. The descended larynx is not uniquely human. *Proc. R. Soc. Lond. B* 268:1669–1675.

FitzGibbon, C. D., and J. H. Fanshawe. 1988. Stotting in Thomson's gazelles: an honest signal of condition. *Behav. Ecol. Sociobiol.* 23:69–74.

Folstad, I., A. M. Hope, A. Karter, and A. Skorping. 1994. Sexually selected color in male sticklebacks: a signal of both parasite exposure and parasite resistance? *Oikos* 69:511–515.

Folstad, I., and A. J. Karter. 1992. Parasites, bright males and the immunocompetence handicap. *Am. Nat.* 139:603–622.

Forstmeier, W., B. Kempenaers, A. Meyer, and B. Leisler. 2002. A novel song parameter correlates with extra-pair paternity and reflects male longevity. *Proc. R. Soc. Lond. B* 269:1479–1485.

Fox, D. L. 1979. *Biochromy: Natural Coloration of Living Things*. Univ. of California Press, Berkeley.

Freeberg, T. M. 1996. Assortative mating in captive cowbirds is predicted by social experience. *Anim. Behav.* 52:1129–1142.

Freeberg, T. M., S. D. Duncan, T. L. Kast, and D. A. Enstrom. 1999. Cultural influences on female mate choice: an experimental test in cowbirds, *Molothrus ater*. *Anim. Behav.* 57:421–426.

Freeberg, T. M., A. P. King, and M. J. West. 2001. Cultural transmission of vocal traditions in cowbirds (*Molothrus ater*) influences courtship patterns and mate preferences. *J. Comp. Psych.* 115:201–211.

Freeman, S. 1987. Male red-winged blackbirds (*Agelaius phoeniceus*) assess the RHP of neighbors by watching contests. *Behav. Ecol. Sociobiol.* 21:307–311.

Frischknecht, M. 1993. The breeding colouration of male three-spined sticklebacks (*Gasterosteus aculeatus*) as an indicator of energy investment in vigour. *Evol. Ecol.* 7:439–450.

Fugle, G. N., and S. I. Rothstein. 1985. Age- and sex-related variation in size and crown plumage brightness in wintering white-crowned sparrows. *J. Field Ornithol.* 56:356–368.

Fugle, G. N., and S. I. Rothstein. 1987. Experiments on the control of deceptive signals of status in white-crowned sparrows. *Auk* 104:188–197.

Fugle, G. N., S. I. Rothstein, C. W. Osenberg, and M. A. McGinley. 1984. Signals of status in wintering white-crowned sparrows, *Zonotrichia leucophrys gambelii*. *Anim. Behav.* 32:86–93.

Garamszegi, L. Z., A. P. Møller, J. Török, G. Michl, P. Péczely, and M. Richard. 2004. Immune challenge mediates vocal communication in a passerine bird: an experiment. *Behav. Ecol.* 15:148–157.

Gebhardt-Henrich, S., and H. Richner. 1998. Causes of growth variation and its conse-quences for fitness. Pp. 324–339 in J. M. Starck and R. E. Ricklefs, eds. *Avian Growth and Development: Evolution within the Altricial-precocial Spectrum*. Oxford Univ. Press, New York.

Geist, V. 1971. *Mountain Sheep: A Study in Behavior and Evolution*. Univ. Chicago Press, Chicago.

Gerhardt, H. C. 1994. The evolution of vocalizations in frogs and toads. *Ann. Rev. Ecol. Syst.* 25:293–324.

Gerhardt, H. C., and F. Huber. 2002. *Acoustic Communication in Insects and Anurans*. Univ. Chicago Press, Chicago.

Getty, T. 1995. Search, discrimination, and selection: mate choice by pied flycatchers. *Am. Nat.* 145:146–154.

Getty, T. 1996. Mate selection by repeated inspection: more on pied flycatchers. *Anim. Behav.* 51:739–745.

Given, M. F. 1987. Vocalizations and acoustic interactions of the carpenter frog, *Rana virgatipes*. *Herpetologica* 43:467–481.

Given, M. F. 1999. Frequency alteration of the advertisement call in the carpenter frog, *Rana virgatipes*. *Herpetologica* 55:304–317.

Godfray, H. C. J. 1991. Signalling of need by offspring to their parents. *Nature* 352:328–330.

Godfray, H. C. J. 1995. Signaling of need between parents and young: parent-offspring conflict and sibling rivalry. *Am. Nat.* 146:1–24.

Godin, J.-G., and H. E. McDonough. 2003. Predator preference for brightly colored males in the guppy: a viability cost for a sexually selected trait. *Behav. Ecol.* 14:194–200.

Gonzalez, G., G. Sorci, A. P. Møller, P. Ninni, C. Haussy, and F. de Lope. 1999. Immu-nocompetence and condition-dependent sexual advertisement in male house spar-rows (*Passer domesticus*). *J. Anim. Ecol.* 68:1225–1234.

Gonzalez, G., G. Sorci, L. C. Smith, and F. de Lope. 2001. Testosterone and sexual signalling in male house sparrows (*Passer domesticus*). *Behav. Ecol. Sociobiol.* 50:557–562.

Gonzalez, G., G. Sorci, L. C. Smith, and F. de Lope. 2002. Social control and physiolog-ical cost of cheating in status signalling male house sparrows (*Passer domesticus*). *Ethology* 108:289–302.

Götmark, F. 1994. Does a novel bright colour patch increase or decrease predation? Red wings reduce predation risk in European blackbirds. *Proc. R. Soc. Lond. B* 256:83–87.

Götmark, F. 1996. Simulating a colour mutation: conspicuous red wings in the Euro-pean blackbird reduce the risk of attacks by sparrowhawks. *Functional Ecology* 10:355–359.

Götmark, F., and M. Ahlström. 1997. Parental preference for red mouth of chicks in a songbird. *Proc. R. Soc. Lond. B* 264:959–962.

Götmark, F., and J. Olsson. 1997. Artificial colour mutation: do red-painted great tits experience increased or decreased predation? *Anim. Behav.* 53:83–91.

Gottlander, K. 1987. Variation in the song rate of the male pied flycatcher *Ficedula hypoleuca*: causes and consequences. *Anim. Behav.* 35:1037–1043.

Grafen, A. 1990a. Sexual selection unhandicapped by the Fisher process. *J. Theor. Biol.* 144:473–516.

Grafen, A. 1990b. Biological signals as handicaps. *J. Theor. Biol.* 144:517–546.

Grafen, A., and R. A. Johnstone. 1993. Why we need ESS signalling theory. *Phil. Trans. R. Soc. Lond. B* 340:245–250.

Grasso, M. J., U. M. Savalli, and R. L. Mumme. 1996. Status signaling in dark-eyed juncos: perceived status of other birds affects dominance interactions. *Condor* 98:636–639.

Greene, E., and T. Meagher. 1998. Red squirrels, *Tamiasciurus hudsonicus*, produce predator-class specific alarm calls. *Anim. Behav.* 55:511–518.

Greenfield, M. D. 2002. *Signalers and Receivers: Mechanisms and Evolution of Arthropod Communication*. Oxford Univ. Press, New York.

Grether, G. F. 2000. Carotenoid limitation and mate preference evolution: a test of the indicator hypothesis in guppies (*Poecilia reticulata*). *Evolution* 54:1712–1724.

Grether, G. F., J. Hudon, and D. F. Millie. 1999. Carotenoid limitation of sexual coloration along an environmental gradient in guppies. *Proc. R. Soc. Lond. B* 266:1317–1322.

Grether, G. F., S. Kasahara, G. R. Kolluru, and E. L. Cooper. 2004. Sex-specific effects of carotenoid intake on the immunological response to allografts in guppies (*Poecilia reticulata*). *Proc. R. Soc. Lond. B* 271:45–49.

Grieg-Smith, P. W. 1982. Song-rates and parental care by individual male stonechats (*Saxicola torquata*). *Anim. Behav.* 30:245–252.

Griesser, M., and J. Ekman. 2004. Nepotistic alarm calling in the Siberian jay, *Perisoreus infaustus*. *Anim. Behav.* 67:933–939.

Gros-Louis, J. 2004. The function of food-associated calls in white-faced capuchin monkeys, *Cebus capucinus*, from the perspective of the signaller. *Anim. Behav.* 67:431–440.

Guilford, T., and M. S. Dawkins. 1991. Receiver psychology and the evolution of animal signals. *Anim. Behav.* 42:1–14.

Guilford, T., and M. S. Dawkins. 1995. What are conventional signals? *Anim. Behav.* 49:1689–1695.

Gyger, M., P. Marler, and R. Pickert. 1987. Semantics of an avian alarm call system: the male domestic fowl, *Gallus domesticus. Behaviour* 102:15–40.

Hamilton, W. D. 1963. The evolution of altruistic behavior. *Am. Nat.* 97:354–356.

Hare, J. F. 1998. Juvenile Richardson's ground squirrels, *Spermophilus richardsonii*, discriminate among individual alarm callers. *Anim. Behav.* 55:451–460.

Hare, J. F., and B. A. Atkins. 2001. The squirrel that cried wolf: reliability detection by juvenile Richardson's ground squirrels (*Spermophilus richardsonii*). *Behav. Ecol. Sociobiol.* 51:108–112.

Harnad, S. 1987. *Categorical Perception: the Groundwork of Cognition*. Cambridge Univ. Press, Cambridge.

Harris, M. P. 1983. Parent-young communication in the puffin *Fratercula arctica. Ibis* 125:109–114.

Haskell, D. 1994. Experimental evidence that nestling begging behaviour incurs a cost due to nest predation. *Proc. Roy. Soc. Lond. B* 257:161–164.

Hasselquist, D. 1998. Polygyny in great reed warblers: A long-term study of factors contributing to male fitness. *Ecology* 79:2376–2390.

Hasselquist, D., S. Bensch, and T. von Schantz. 1996. Correlation between male song repertoire, extra-pair paternity and offspring survival in the great reed warbler. *Nature* 381:229–232.

Hasselquist, D., J. A. Marsh, P. W. Sherman, and J. C. Wingfield. 1999. Is avian humoral immunocompetence suppressed by testosterone? *Behav. Ecol. Sociobiol.* 45:167–175.

Hauber, M. E., and P. W. Sherman. 1998. Nepotism and marmot alarm calling. *Anim. Behav.* 56:1049–1052.

Hauser, M. D. 1996. *The Evolution of Communication*. MIT Press, Cambridge, MA.

Hauser, M. D., and P. Marler. 1993a. Food-associated calls in rhesus macaques (*Macaca mulatta*): II. Costs and benefits of call production and suppression. *Behav. Ecol.* 4:206–212.

Hauser, M. D., and P. Marler. 1993b. Food-associated calls in rhesus macaques (*Macaca mulatta*): I. Socioecological factors. *Behav. Ecol.* 4:194–205.

Hauser, M. D., P. Teixidor, L. Fields, and R. Flaherty. 1993. Food-elicited calls in chimpanzees: effects of food quantity and divisibility. *Anim. Behav.* 45:817–819.

Hauser, M. D., and R. W. Wrangham. 1987. Manipulation of food calls in captive chimpanzees. *Folia Primatol.* 48:207–210.

Hazlett, B. A., and W. H. Bossert. 1965. A statistical analysis of the aggressive communication systems of some hermit crabs. *Anim. Behav.* 13:357–373.

Hedrick, A. V. 1986. Female preferences for male calling bout duration in a field cricket. *Behav. Ecol. Sociobiol.* 19:73–77.

Hegner, R. E., and J. C. Wingfield. 1986. Behavioral and endocrine correlates of multiple brooding in the semicolonial house sparrow *Passer domesticus*. I. Males. *Horm. Behav.* 20:294–312.

Hein, W. K., D. F. Westneat, and J. P. Poston. 2003. Sex of opponent influences response to a potential status signal in house sparrows. *Anim. Behav.* 65:1211–1221.

Heinrich, B. 1988. Winter foraging at carcasses by three sympatric corvids, with emphasis on recruitment by the raven, *Corvus corax. Behav. Ecol. Sociobiol.* 23:141–156.

Heinrich, B., and J. M. Marzluff. 1991. Do common ravens yell because they want to attract others? *Behav. Ecol. Sociobiol.* 28:13–21.

Herb, B. M., S. A. Biron, and M. R. Kidd. 2003. Courtship by subordinate male Siamese fighting fish, *Betta splendens*: their response to eavesdropping and naïve females. *Behaviour* 140:71–78.

Hiebert, S. M., P. K. Stoddard, and P. Arcese. 1989. Repertoire size, territory acquisition and reproductive success in the song sparrow. *Anim. Behavior* 37:266–273.

Hill, G. E. 1990. Female house finches prefer colourful males: sexual selection for a condition-dependent trait. *Anim. Behav.* 40:563–572.

Hill, G. E. 1991. Plumage coloration is a sexually selected indicator of male quality. *Nature* 350:337–339.

Hill, G. E. 1992. Proximate basis of variation in carotenoid pigmentation in male house finches. *Auk* 109:1–12.

Hill, G. E. 1994. House finches are what they eat: a reply to Hudon. *Auk* 111:221–225.

Hill, G. E. 1995. Seasonal variation in circulating carotenoid pigments in the house finch. *Auk* 112:1057–1061.

Hill, G. E. 1999. Is there an immunological cost to carotenoid-based ornamental coloration? *Am. Nat.* 154:589–595.

Hill, G. E. 2000. Energetic constraints on expression of carotenoid-based plumage coloration. *J. Avian Biol.* 31:559–566.

Hill, G. E. 2002. *A Red Bird in a Brown Bag*. Oxford Univ. Press, Oxford.

Hill, G. E., C. Y. Inouye, and R. Montgomerie. 2002. Dietary carotenoids predict plumage coloration in wild house finches. *Proc. R. Soc. Lond. B* 269:1119–1124.

Hill, G. E., P. M. Nolan, and A. M. Stoehr. 1999. Pairing success relative to male plumage redness and pigment symmetry in the house finch: temporal and geographic constancy. *Behav. Ecol.* 10:48–53.

Hinde, R. A. 1981. Animal signals: ethological and games-theory approaches are not incompatible. *Anim. Behav.* 29:535–542.

Hirth, D. H., and D. R. McCullough. 1977. Evolution of alarm signals in ungulates with special reference to white-tailed deer. *Am. Nat.* 111:31–42.

Hoese, W. J., J. Podos, N. C. Boetticher, and S. Nowicki. 2000. Vocal tract function in birdsong production: experimental manipulation of beak movements. *J. Exp. Biol.* 203:1845–1855.

Hogstad, O. 1987. It is expensive to be dominant. *Auk* 104:333–336.

Hoi-Leitner, M., H. Nechtelberger, and J. Dittami. 1993. The relationship between individual differences in male song frequency and parental care in blackcaps. *Behaviour* 126:1–12.

Hoi-Leitner, M., H. Nechtelberger, and H. Hoi. 1995. Song rate as a signal for nest site quality in blackcaps (*Sylvia atricapilla*). *Behav. Ecol. Sociobiol.* 37:399–405.

Holberton, R. L., K. P. Able, and J. C. Wingfield. 1989. Status signalling in dark-eyed juncos, *Junco hyemalis*: plumage manipulations and hormonal correlates of dominance. *Anim. Behav.* 37:681–689.

Hoogland, J. L. 1983. Nepotism and alarm calling in the black-tailed prairie dog (*Cynomys ludovicianus*). *Anim. Behav.* 31:472–479.

Hoogland, J. L. 1996. Why do Gunnison's prairie dogs give anti-predator calls? *Anim. Behav.* 51:871–880.

Hõrak, P., I. Ots, H. Vellau, C. Spottiswoode, and A. P. Møller. 2001. Carotenoid-based plumage coloration reflects hemoparasite infection and local survival in breeding great tits. *Oecologia* 126:166–173.

Houde, A. E. 1987. Mate choice based upon naturally occurring color-pattern variation in a guppy population. *Evolution* 41:1–10.

Houde, A. E., and A. J. Torio. 1992. Effect of parasitic infection on male color pattern and female choice in guppies. *Behav. Ecol.* 3:346–351.

Houtman, A. M. 1992. Female zebra finches choose extra-pair copulations with genetically attractive males. *Proc. R. Soc. Lond. B* 248:3–6.

Howard, R. D. 1974. The influence of sexual selection and interspecific competition on mockingbird song (*Mimus polyglottos*). *Evolution* 28:428–438.

Howard, R. D. 1978. The evolution of mating strategies in bullfrogs, *Rana catesbeiana*. *Evolution* 32:850–871.

Howard, R. D., and J. R. Young. 1998. Individual variation in male vocal traits and female mating preferences in *Bufo americanus*. *Anim. Behav.* 55:1165–1179.

Hsu, Y., and L. L. Wolf. 1999. The winner and loser effect: integrating multiple social experiences. *Anim. Behav.* 57:903–910.

Hsu, Y., and L. L. Wolf. 2001. The winner and loser effect: what fighting behaviours are influenced? *Anim. Behav.* 61:777–786.

Hudon, J. 1994. Showiness, carotenoids, and captivity: a comment on Hill (1992). *Auk* 111:218–221.

Hughes, M. 1996. Size assessment via a visual signal in snapping shrimp. *Behav. Ecol. Sociobiol.* 38:51–57.

Hughes, M. 2000. Deception with honest signals: signal residuals and signal function in snapping shrimp. *Behav. Ecol.* 11:614–623.

Hughes, M., S. Nowicki, W. A. Searcy, and S. Peters. 1998. Song-type sharing in song sparrows: implications for repertoire function and song learning. *Behav. Ecol. Sociobiol.* 42:437–446.

Hultsch, H., and D. Todt. 1982. Temporal performance roles during vocal interactions in nightingales (*Luscinia megarhynchos* B.). *Behav. Ecol. Sociobiol.* 11:253–260.

Hurd, P. L. 1997. Is signalling of fighting ability costlier for weaker individuals? *J. Theor. Biol.* 184:83–88.

Iacovides, S., and R. M. Evans. 1998. Begging as graded signals of need for food in young ring-billed gulls. *Anim. Behav.* 56:79–85.

Immelmann, K. 1969. Song development in the zebra finch and other estrildid finches. Pp. 61–74 in R. A. Hinde, ed., *Bird Vocalizations*. Cambridge Univ. Press, Cambridge.

Iwasa, Y., A. Pomiankowski, and S. Nee. 1991. The evolution of costly mate preferences. II. The "handicap" principle. *Evolution* 45:1431–1442.

Jackson, W. M. 1991. Why do winners keep winning? *Behav. Ecol. Sociobiol.* 28:271–276.

Johnsson, J. I., and A. Åkerman. 1998. Watch and learn: preview of the fighting ability of opponents alters contest behaviour in rainbow trout. *Anim. Behav.* 56:771–776.

Johnstone, R. A. 1997. The evolution of animal signals. Pp. 155–178 in J. R. Krebs and N. B. Davies, eds., *Behavioural Ecology*. Blackwell, Oxford.

Johnstone, R. A. 1998. Conspiratorial whispers and conspicuous displays: games of signal detection. *Evolution* 52:1554–1563.

Johnstone, R. A. 1999. Signaling of need, sibling competition, and the cost of honesty. *Proc. Natl. Acad. Sci. USA* 96:12644–12649.

Johnstone, R. A. 2001. Eavesdropping and animal conflict. *Proc. Natl. Acad. Sci. USA* 98:9177–9180.

Johnstone, R. A. 2004. Begging and sibling competition: how should offspring respond to their rivals? *Am. Nat.* 163:388–406.

Johnstone, R. A., and A. Grafen. 1992. The continuous Sir Philip Sidney Game: a simple model of biological signalling. *J. Theor. Biol.* 156:215–234.

Johnstone, R. A., and A. Grafen. 1993. Dishonesty and the handicap principle. *Anim. Behav.* 46:759–764.

Johnstone, R. A., and K. Norris. 1993. Badges of status and the cost of aggression. *Behav. Ecol. Sociobiol.* 32:127–134.

Jones, T. M., R. J. Quinnell, and A. Balmford. 1998. Fisherian flies: benefits of female choice in a lekking sandfly. *Proc. Roy. Soc. Lond. B* 265:1651–1657.

Kacelnik, A., P. A. Cotton, L. Stirling, and J. Wright. 1995. Food allocation among nestling starlings: sibling competition and the scope of parental choice. *Proc. Roy. Soc. Lond. B* 259:259–263.

Kedar, H., M. A. Rodríguez-Gironés, S. Yedvab, D. W. Winkler, and A. Lotem. 2000. Experimental evidence for offspring learning in parent-offspring communication. *Proc. R. Soc. Lond. B* 267:1723–1727.

Keys, G. C., and S. I. Rothstein. 1991. Benefits and costs of dominance and subordinance in white-crowned sparrows and the paradox of status signalling. *Anim. Behav.* 42:899–912.

Kilner, R., and R. A. Johnstone. 1997. Begging the question: are offspring solicitation behaviours signals of need? *Trends Ecol. Evol.* 12:11–15.

Kilner, R. M. 1997. Mouth colour is a reliable signal of need in begging canary nestlings. *Proc. R. Soc. Lond. B* 264:963–968.

Kilner, R. M. 2001. A growth cost of begging in captive canary chicks. *Proc. Natl. Acad. Sci. USA* 98:11394–11398.

Kilner, R. M., D. G. Noble, and N. B. Davies. 1999. Signals of need in parent-offspring communication and their exploitation by the common cuckoo. *Nature* 397:667–672.

King, A. P., and M. J. West. 1977. Species identification in the North American cowbird: appropriate responses to abnormal song. *Science* 195:1002–1004.

King, A. P., M. J. West, and D. H. Eastzer. 1980. Song structure and song development as potential contributors to reproductive isolation in cowbirds (*Molothrus ater*). *J. Comp. Physiol. Psychol.* 94:1028–1039.

Kirkpatrick, M. 1982. Sexual selection and the evolution of female choice. *Evolution* 36:1–12.

Kirkpatrick, M. 1985. Evolution of female choice and male parental investment in polygynous species: the demise of the "sexy son." *Am. Nat.* 125:788–810.

Kirkpatrick, M., and M. J. Ryan. 1991. The evolution of mating preferences and the paradox of the lek. *Nature* 350:33–38.

Kirn, J. R., R. P. Clower, D. E. Kroodsma, and T. J. DeVoogd. 1989. Song-related brain regions in the red-winged blackbird are affected by sex and season but not repertoire size. *J. Neurobiol.* 20:139–163.

Klump, G. M., E. Kretzschmar, and E. Curio. 1986. The hearing of an avian predator and its avian prey. *Behav. Ecol. Sociobiol.* 18:317–323.

Klump, G. M., and M. D. Shalter. 1984. Acoustic behaviour of birds and mammals in the predator context. *Z. Tierpsychol.* 66:189–226.

Knowlton, N., and B. D. Keller. 1982. Symmetric fights as a measure of escalation potential in a symbiotic, territorial snapping shrimp. *Behav. Ecol. Sociobiol.* 10:289–292.

Kodric-Brown, A. 1985. Female preference and sexual selection for male coloration in the guppy (*Poecilia reticulata*). *Behav. Ecol. Sociobiol.* 17:199–205.

Kodric-Brown, A. 1989. Dietary carotenoids and male mating success in the guppy: an environmental component to female choice. *Behav. Ecol. Sociobiol.* 25:393–401.

Kokko, H. 1997. Evolutionarily stable strategies of age-dependent sexual advertisement. *Behav. Ecol. Sociobiol.* 41:99–107.

Kokko, H. 1998. Should advertising parental care be honest? *Proc. R. Soc. Lond. B* 265:1871–1878.

Kotiaho, J. S. 2000. Testing the assumptions of conditional handicap theory: costs and condition dependence of a sexually selected trait. *Behav. Ecol. Sociobiol.* 48:188–194.

Kotiaho, J. S. 2001. Costs of sexual traits: a mismatch between theoretical considerations and empirical evidence. *Biol. Rev.* 76:365–376.

Kotiaho, J. S., R. V. Alatalo, J. Mappes, and S. Parri. 1996. Sexual selection in a wolf spider: male drumming activity, body size and viability. *Evolution* 50:1977–1981.

Kotiaho, J. S., R. V. Alatalo, J. Mappes, S. Parri, and A. Rivero. 1998. Male mating success and risk of predation in a wolf spider: a balance between sexual and natural selection? *J. Anim. Ecol.* 67:287–291.

Kramer, H. G., and R. E. Lemon. 1983. Dynamics of territorial singing between neighboring song sparrows (*Melospiza melodia*). *Behaviour* 85:198–223.

Krebs, J., R. Ashcroft, and M. Webber. 1978. Song repertoires and territory defence in the great tit. *Nature* 271:539–542.

Kroodsma, D. E., M. C. Baker, L. F. Baptista, and L. Petrinovich. 1985. Vocal "dialects" in Nuttall's white-crowned sparrow. *Curr. Ornithol.* 2:103–133.

Lachmann, M., and C. T. Bergstrom. 1998. Signalling among relatives. II. Beyond the Tower of Babel. *Theor. Pop. Biol.* 54:146–160.

Lack, D. 1966. *Population Studies of Birds*. Oxford Univ. Press, Oxford.

Lahti, K. 1998. Social dominance and survival in flocking passerine birds: A review with an emphasis on the willow tit *Parus montanus*. *Ornis Fennica* 75:1–17.

Lambrechts, M., and A. A. Dhondt. 1988. The anti-exhaustion hypothesis: a new hypothesis to explain song performance and song switching in the great tit. *Anim. Behav.* 36:327–334.

Lampe, H. M., and Y. O. Espmark. 1994. Song structure reflects male quality in pied flycatchers, *Ficedula hypoleuca. Anim. Behav.* 47:869–876.

Lampe, H. M., and G.-P. Sætre. 1995. Female pied flycatchers prefer males with larger song repertoires. *Proc. R. Soc. Lond. B* 262:163–167.

Lande, R. 1981. Models of speciation by sexual selection on polygenic traits. *Proc. Natl. Acad. Sci. USA* 78:3721–3725.

Langemann, U., J. P. Tavares, T. M. Peake, and P. K. McGregor. 2000. Response of great tits to escalating patterns of playback. *Behaviour* 137:451–471.

Leech, S. M., and M. L. Leonard. 1996. Is there an energetic cost to begging in nestling tree swallows (*Tachycineta bicolor*)? *Proc. Roy. Soc. Lond. B* 263:983–987.

Leech, S. M., and M. L. Leonard. 1997. Begging and the risk of predation in nestling birds. *Behav. Ecol.* 8:644–646.

Leger, D. W., and D. H. Owings. 1978. Responses to alarm calls by California ground squirrels: effects of call structure and maternal status. *Behav. Ecol. Sociobiol.* 3:177–186.

Leger, D. W., D. H. Owings, and L. M. Boal. 1979. Contextual information and differential responses to alarm whistles in California ground squirrels. *Z. Tierpsychol.* 49:142–155.

Leger, D. W., D. H. Owings, and D. L. Gelfand. 1980. Single-note vocalizations of California ground squirrels: graded signals and situation-specificity of predator and socially evoked calls. *Z. Tierpsychol.* 52:227–246.

Lemel, J., and K. Wallin. 1993. Status signalling, motivational condition and dominance: an experimental study in the great tit, *Parus major* L. *Anim. Behav.* 45:549–558.

Leonard, M. L., and A. G. Horn. 1998. Need and nestmates affect begging in tree swallows. *Behav. Ecol. Sociobiol.* 42:431–436.

Leonard, M. L., A. G. Horn, and E. Parks. 2003a. The role of posturing and calling in the begging display of nestling birds. *Behav. Ecol. Sociobiol.* 54:188–193.

Leonard, M. L., A. G. Horn, and J. Porter. 2003b. Does begging affect growth in nestling tree swallows, *Tachycineta bicolor? Behav. Ecol. Sociobiol.* 54:573–577.

Levitsky, D. A., and B. J. Strupp. 1995. Malnutrition and the brain: changing concepts, changing concerns. *J. Nutrit.* 125:2212S-2220S.

Lind, H., T. Dabelsteen, and P. K. McGregor. 1996. Female great tits can identify mates by song. *Anim. Behav.* 52:667–671.

Lindström, K. M., D. Krakower, J. O. Lundström, and B. Silverin. 2001. The effects of testosterone on a viral infection in greenfinches (*Carduelis chloris*): an experimental test of the immunocompetence-handicap hypothesis. *Proc. R. Soc. Lond. B* 268:207–211.

Litovich, E., and H. W. Power. 1992. Parent-offspring conflict and its resolution in the European starling. *Ornithol. Monogr.* 47:1–71.

Little, W., H. W. Fowler, J. Coulson, and C. T. Onions. 1964. *The Oxford Universal Dictionary.* Oxford Univ. Press, Oxford.

Lloyd, J. E. 1965. Aggressive mimicry in *Photuris*: firefly femmes fatales. *Science* 149:653–654.

Lloyd, J. E. 1986. Firefly communication and deception: "Oh, what a tangled web!" Pp. 113–128 in R. W. Mitchell and R. W. Thompson, eds., *Deception: Perspectives on Human and Nonhuman Deceit.* SUNY Press, Albany.

Lopez, P. T., P. M. Narins, E. R. Lewis, and S. W. Moore. 1988. Acoustically induced call modification in the white-lipped frog, *Leptodactylus albilabris. Anim. Behav.* 36:1295–1308.

Lozano, G. A. 1994. Carotenoid, parasites, and sexual selection. *Oikos* 70:309–311.

MacDougall-Shackleton, E. A., E. P. Derryberry, and T. P. Hahn. 2002. Nonlocal male mountain white-crowned sparrows have lower paternity and higher parasite loads than males singing local dialect. *Behav. Ecol.* 13:682–689.

MacDougall-Shackleton, E. A., and S. A. MacDougall-Shackleton. 2001. Cultural and genetic evolution in mountain white-crowned sparrows: song dialects are associated with population structure. *Evolution* 55:2568–2575.

Macnair, M. R., and G. A. Parker. 1979. Models of parent-offspring conflict. III. Intra-brood conflict. *Anim. Behav.* 27:1202–1209.

Manser, M. B. 2001. The acoustic structure of suricates' alarm calls varies with predator type and the level of response urgency. *Proc. R. Soc. Lond. B* 268:2315–2324.

Manser, M. B., M. B. Bell, and L. B. Fletcher. 2001. The information that receivers extract from alarm calls in suricates. *Proc. R. Soc. Lond. B* 268:2485–2491.

Mappes, J., R. V. Alatalo, J. Kotiaho, and S. Parri. 1996. Viability costs of condition-dependent sexual male display in a drumming wolf spider. *Proc. R. Soc. Lond. B* 263:785–789.

Marler, P. 1955. Characteristics of some animal calls. *Nature* 176:6–8.

Marler, P. 1968. Visual systems. Pp. 103–126 in T. A. Sebeok, ed., *Animal Communication.* Indiana Univ. Press, Bloomington, IN.

Marler, P. 1970. A comparative approach to vocal learning: song development in white-crowned sparrows. *J. Comp. Phys. Psych.* 71:1–25.

Marler, P. 1977. The evolution of communication. Pp. 45–70 in T. A. Sebeok, ed., *How Animals Communicate.* Indiana Univ. Press, Bloomington, IN.

Marler, P., A. Dufty, and R. Pickert. 1986a. Vocal communication in the domestic chicken: I. Does a sender communicate information about the quality of a food referent to a receiver? *Anim. Behav.* 34:188–193.

Marler, P., A. Dufty, and R. Pickert. 1986b. Vocal communication in the domestic chicken: II. Is a sender sensitive to the presence and nature of a receiver? *Anim. Behav.* 34:194–198.

Marler, P., and W. J. Hamilton. 1966. *Mechanisms of Animal Behavior.* Wiley, New York.

Marler, P., and S. Peters. 1987. A sensitive period for song acquisition in the song sparrow, *Melospiza melodia*: A case of age-limited learning. *Ethology* 76:89–100.

Marler, P., and S. Peters. 1988. Sensitive periods for song acquisition from tape recordings and live tutors in the swamp sparrow, *Melospiza georgiana. Ethology* 77:76–84.

Marler, P., and M. Tamura. 1962. Song "dialects" in three populations of white-crowned sparrows. *Condor* 64:368–377.

Martin, T. E. 1987. Food as a limit on breeding birds: a life-history perspective. *Ann. Rev. Ecol. Syst.* 18:453–487.

Martin, W. F. 1971. Mechanics of sound production in toads of the genus *Bufo*: passive elements. *J. Exp. Zool.* 176:273–294.

Martin, W. F. 1972. Evolution of vocalizations in the genus *Bufo*. Pp. 279–309 in W. F. Blair, ed. *Evolution in the Genus Bufo*. Univ. Texas Press, Austin.

Masataka, N. 1994. Lack of correlation between body size and frequency of vocalizations in young female Japanese macaques (*Macaca fuscata*). *Folia Primatol.* 63:115–118.

Mateos, C., and J. Carranza. 1995. Female choice for morphological features of male ring-necked pheasants. *Anim. Behav.* 49:737–748.

Matos, R. J., and P. K. McGregor. 2002. The effect of the sex of an audience on male-male displays of Siamese fighting fish (*Betta splendens*). *Behaviour* 139:1211–1221.

Matos, R. J., T. M. Peake, and P. K. McGregor. 2003. Timing of presentation of an audience: aggressive priming and audience effects in male displays of Siamese fighting fish (*Betta splendens*). *Behav. Process.* 63:53–61.

Matsuoka, S. 1980. Pseudo warning call in titmice. *Tori* 29:87–90.

Maurer, G., R. D. Magrath, M. L. Leonard, A. G. Horn, and C. Donnelly. 2003. Begging to differ: scrubwren nestlings beg to alarm calls and vocalize when parents are absent. *Anim. Behav.* 65:1045–1055.

Maynard Smith, J. 1965. The evolution of alarm calls. *Am. Nat.* 99:59–63.

Maynard Smith, J. 1974. The theory of games and the evolution of animal conflicts. *J. Theor. Biol.* 47:209–221.

Maynard Smith, J. 1976a. Group selection. *Quart. Rev. Biol.* 51:277–283.

Maynard Smith, J. 1976b. Sexual selection and the handicap principle. *J. Theor. Biol.* 57:239–242.

Maynard Smith, J. 1979. Game theory and the evolution of behaviour. *Proc. R. Soc. Lond. B* 205:475–488.

Maynard Smith, J. 1982. *Evolution and the Theory of Games*. Cambridge Univ. Press, Cambridge.

Maynard Smith, J. 1985. Sexual selection, handicaps and true fitness. *J. Theor. Biol.* 115:1–8.

Maynard Smith, J. 1991a. Honest signalling: the Philip Sidney game. *Anim. Behav.* 42:1034–1035.

Maynard Smith, J. 1991b. Theories of sexual selection. *Trends. Ecol. Evol.* 6:146–151.

Maynard Smith, J., and D. Harper. 2003. *Animal Signals*. Oxford Univ. Press, Oxford.

Maynard Smith, J., and D. G. C. Harper. 1988. The evolution of aggression: can selection generate variability? *Phil. Trans. R. Soc. Lond. B* 319:557–570.

Maynard Smith, J., and G. A. Parker. 1976. The logic of asymmetric contests. *Anim. Behav.* 24:159–175.

Maynard Smith, J., and G. R. Price. 1973. The logic of animal conflict. *Nature* 246:15–18.

McCarty, J. P. 1996. The energetic cost of begging in nestling passerines. *Auk* 113: 178–188.

McGraw, K. J., and D. R. Ardia. 2003. Carotenoids, immunocompetence, and the information content of sexual colors: an experimental test. *Am. Nat.* 162:704–712.

McGraw, K. J., and G. E. Hill. 2000. Carotenoid-based ornamentation and status signaling in the house finch. *Behav. Ecol.* 11:520–527.

McGraw, K. J., E. A. Mackillop, J. Dale, and M. E. Hauber. 2002. Different colors reveal different information: how nutritional stress affects the expression of melanin- and structurally based ornamental plumage. *J. Exp. Biol.* 205:3747–3755.

McGregor, P. K. 1980. Song dialects in the corn bunting (*Emberiza calandra*). *Z. Tierpsychol.* 54:285–297.

McGregor, P. K. 1993. Signalling in territorial systems: a context for individual identification, ranging and eavesdropping. *Phil. Trans. R. Soc. Lond. B* 340:237–244.

McGregor, P. K., and T. Dabelsteen. 1996. Communication networks. Pp. 409–425 in D. E. Kroodsma and E. H. Miller, eds.. *Ecology and Evolution of Acoustic Communication in Birds*. Cornell Univ. Press, Ithaca, NY.

McGregor, P. K., T. Dabelsteen, M. Shepherd, and S. B. Pedersen. 1992. The signal value of matched singing in great tits: evidence from interactive playback experiments. *Anim. Behav.* 43:987–998.

McGregor, P. K., J. R. Krebs, and C. M. Perrins. 1981. Song repertoires and lifetime reproductive success in the great tit (*Parus major*). *Am. Nat.* 118:149–159.

McGregor, P. K., K. Otter, and T. M. Peake. 2000. Communication networks: receiver and signaller perspectives. Pp. 329–340 in Y. Espmark, T. Amundsen, and G. Rosenqvist, eds., *Animal Signals: Signalling and Signal Design in Animal Communication*. Tapir Academic Press, Trondheim, Norway.

McGregor, P. K., and T. M. Peake. 2000. Communication networks: social environments for receiving and signalling behaviour. *Acta. Ethol.* 2:71–81.

McGregor, P. K., T. M. Peake, and H. M. Lampe. 2001. Fighting fish *Betta splendens* extract relative information from apparent interactions: what happens when what you see is not what you get. *Anim. Behav.* 62:1059–1065.

Melchior, H. R. 1971. Characteristics of arctic ground squirrel alarm calls. *Oecologia* 7:184–190.

Mennill, D. J., P. T. Boag, and L. M. Ratcliffe. 2003. The reproductive choices of eavesdropping female black-capped chickadees, *Poecile atricapillus*. *Naturwissenschaften* 90:577–582.

Mennill, D. J., and L. M. Ratcliffe. 2004. Do male black-capped chickadees eavesdrop on song contests? A multi-speaker playback experiment. *Behaviour* 141:125–139.

Mennill, D. J., L. M. Ratcliffe, and P. T. Boag. 2002. Female eavesdropping on male song contests in songbirds. *Science* 296:873.

Milinski, M., and T. C. M. Bakker. 1990. Female sticklebacks use male coloration in mate choice and hence avoid parasitized males. *Nature* 344:330–333.

Miller, G. R., and J. B. Stiff. 1993. *Deceptive Communication*. Sage Publications, Newbury Park, CA.

Mitchell, R. W. 1986. A framework for discussing deception. Pp. 3–40 in R. W. Mitchell and N. S. Thompson, eds., *Deception: Perspectives on Human and Nonhuman Deceit*. SUNY Press, Albany.

Møller, A. P. 1987a. Variation in badge size in male house sparrows *Passer domesticus*: evidence for status signalling. *Anim. Behav.* 35:1637–1644.

Møller, A. P. 1987b. Social control of deception among status signalliing house sparrows *Passer domesticus*. *Behav. Ecol. Sociobiol.* 20:307–311.

Møller, A. P. 1988a. False alarm calls as a means of resource usurpation in the great tit *Parus major*. *Ethology* 79:25–30.

Møller, A. P. 1988b. Female choice selects for male sexual tail ornaments in the monogamous swallow. *Nature* 332:640–642.

Møller, A. P. 1989a. Deceptive use of alarm calls by male swallows, *Hirundo rustica*: a new paternity guard. *Behav. Ecol.* 1:1–6.

Møller, A. P. 1989b. Viability costs of male tail ornaments in a swallow. *Nature* 339:132–135.

Møller, A. P. 1990a. Effects of a haematophagous mite on the barn swallow (*Hirundo rustica*): a test of the Hamilton and Zuk hypothesis. *Evolution* 44:771–784.

Møller, A. P. 1990b. Male tail length and female mate choice in the monogamous swallow *Hirundo rustica*. *Anim. Behav.* 39:458–465.

Møller, A. P. 1991a. Parasite load reduces song output in a passerine bird. *Anim. Behav.* 41:723–730.

Møller, A. P. 1991b. Parasites, sexual ornaments, and mate choice in the barn swallow. Pp. 328–348 in J. Loye and M. Zuk, eds., *Bird-Parasite Interactions: Ecology, Evolution, and Behaviour*. Oxford Univ. Press, Oxford.

Møller, A. P. 1992. Sexual selection in the monogamous swallow (*Hirundo rustica*). II. Mechanisms of intersexual selection. *J. Evol. Biol.* 5:603–624.

Møller, A. P. 1994a. *Sexual Selection and the Barn Swallow*. Oxford Univ. Press, Oxford.

Møller, A. P. 1994b. Male ornament size as a reliable cue to enhanced offspring viability in the barn swallow. *Proc. Natl. Acad. Sci. USA* 91:6929–6932.

Møller, A. P., and F. de Lope. 1994. Differential costs of a secondary sexual character: an experimental test of the handicap principle. *Evolution* 48:1676–1683.

Møller, A. P., and H. Tegelström. 1997. Extra-pair paternity and tail ornamentation in the barn swallow *Hirundo rustica*. *Behav. Ecol. Sociobiol.* 41:353–360.

Mondloch, C. J. 1995. Chick hunger and begging affect parental allocation of feedings in pigeons. *Anim. Behav.* 49:601–613.

Moodie, G. E. E. 1972. Predation, natural selection and adaptation in an unusual threespine stickleback. *Heredity* 28:155–167.

Morton, E. S. 1982. Grading, discreteness, redundancy, and motivation-structural rules. Pp. 183–212 in D. E. Kroodsma and E. H. Miller, eds., *Acoustic Communication in Birds*. vol. 1. Academic Press, New York.

Mountjoy, D. J., and R. E. Lemon. 1996. Female choice for complex song in the European starling: a field experiment. *Behav. Ecol. Sociobiol.* 38:65–71.

Mountjoy, D. J., and R. E. Lemon. 1997. Male song complexity and parental care in the European starling. *Behaviour* 134:661–675.

Muller, R. E., and D. G. Smith. 1978. Parent-offspring interactions in zebra finches. *Auk* 95:485–495.

Munn, C. A. 1986. Birds that 'cry wolf'. *Nature* 319:143–145.

Myrberg, A. A. 1981. Sound communication and interception in fishes. Pp. 395–425 in W. N. Tavolga, A. N. Popper, and R. R. Fay, eds., *Hearing and Sound Communication in Fishes*. Springer-Verlag, Heidelberg.

Naguib, M., C. Fichtel, and D. Todt. 1999. Nightingales respond more strongly to vocal leaders of simulated dyadic interactions. *Proc. R. Soc. Lond. B* 266:537–542.

Naguib, M., and D. Todt. 1997. Effects of dyadic vocal interactions on other conspecific receivers in nightingales. *Anim. Behav.* 54:1535–1543.

Navara, K. J., and G. E. Hill. 2003. Dietary carotenoid pigments and immune function in a songbird with extensive carotenoid-based plumage coloration. *Behav. Ecol.* 14:909–916.

Nelson, D. A. 1984. Communication of intentions in agonistic contexts by the pigeon guillemot, *Cepphus columba*. *Behaviour* 88:145–189.

Nelson, D. A. 1998a. External validity and experimental design: the sensitive phase for song learning. *Anim. Behav.* 56:487–491.

Nelson, D. A. 1998b. Geographic variation in song of Gambel's white-crowned sparrow. *Behaviour* 135:321–342.

Nolan, P. M., G. E. Hill, and A. M. Stoehr. 1998. Sex, size, and plumage redness predict house finch survival in an epidemic. *Proc. R. Soc. Lond. B* 265:961–965.

Nordby, J. C., S. E. Campbell, and M. D. Beecher. 2002. Adult song sparrows do not alter their song repertoires. *Ethology* 108:39–50.

Nordeen, E. J., and K. W. Nordeen. 1988. Sex and regional differences in the incorporation of neurons born during song learning in zebra finches. *J. Neurosci.* 8:2869–2874.

Nordeen, K. W., P. Marler, and E. J. Nordeen. 1989. Addition of song-related neurons in swamp sparrows coincides with memorization, not production, of learned songs. *J. Neurobiol.* 20:651–661.

Nottebohm, F. 1969. The song of the chingolo, *Zonotrichia capensis*, in Argentina: description and evaluation of a system of dialects. *Condor* 71:299–315.

Nottebohm, F. 1972. The origins of vocal learning. *Am. Nat.* 106:116–140.

Nottebohm, F., S. Kasparian, and C. Pandazis. 1981. Brain space for a learned task. *Brain Res.* 213:99–109.

Nottebohm, F., T. M. Stokes, and C. M. Leonard. 1976. Central control of song in the canary, *Serinus canarius*. *J. Comp. Neurol.* 165:457–486.

Nowak, M. A., and K. Sigmund. 1998. Evolution of indirect reciprocity by image scoring. *Nature* 393:573–577.

Nowicki, S. 1987. Vocal tract resonances in oscine bird sound production: evidence from birdsongs in a helium atmosphere. *Nature* 325:53–55.

Nowicki, S., D. Hasselquist, S. Bensch, and S. Peters. 2000. Nestling growth and song repertoire size in great reed warblers: evidence for song learning as an indicator mechanism in mate choice. *Proc. R. Soc. Lond. B* 267:2419–2424.

Nowicki, S., S. Peters, and J. Podos. 1998. Song learning, early nutrition and sexual selection in songbirds. *Am. Zool.* 38:179–190.

Nowicki, S., and W. A. Searcy. 2004. Song function and the evolution of female preferences: why birds sing and why brains matter. Pp. 704–723 in H. P. Ziegler and

P. Marler, ed., *The Behavioral Neurobiology of Bird Song*. New York Academy of Sciences, New York.

Nowicki, S., W. A. Searcy, M. Hughes, and J. Podos. 2001. The evolution of bird song: male and female response to song innovation in swamp sparrows. *Anim. Behav.* 62:1189–1195.

Nowicki, S., W. A. Searcy, and S. Peters. 2002a. Brain development, song learning and mate choice in birds: a review and experimental test of the "nutritional stress hypothesis." *J. Comp. Phys. A* 188:1003–1014.

Nowicki, S., W. A. Searcy, and S. Peters. 2002b. Quality of song learning affects female response to male bird song. *Proc. R. Soc. Lond. B* 269:1949–1954.

Nowicki, S., M. Westneat, and W. Hoese. 1992. Birdsong: motor function and the evolution of communication. *Seminars in Neuroscience* 4:385–390.

O'Connor, R. J. 1984. *The Growth and Development of Birds*. John Wiley & Sons, Chichester.

Oberweger, K., and F. Goller. 2001. The metabolic cost of birdsong production. *J. Exp. Biol.* 204:3379–3388.

Oldham, R. S., and H. C. Gerhardt. 1975. Behavioral isolating mechanisms of the treefrogs *Hyla cinerea* and *H. gratiosa*. *Copeia* 1975:223–231.

Oliveira, R. F., P. K. McGregor, and C. Latruffe. 1998. Know thine enemy: fighting fish gather information from observing conspecific interactions. *Proc. R. Soc. Lond. B* 265:1045–1049.

Olson, V. A., and I. P. F. Owens. 1998. Costly sexual signals: are carotenoids rare, risky or required? *Trends Ecol. Evol.* 13:510–514.

Ophir, A. G., and B. G. Galef. 2003. Female Japanese quail that 'eavesdrop' on fighting males prefer losers to winners. *Anim. Behav.* 66:399–407.

Otte, D. 1974. Effects and functions in the evolution of signaling systems. *Ann. Rev. Ecol. Syst.* 5:385–417.

Otter, K., P. K. McGregor, A. M. R. Terry, F. R. L. Burford, T. M. Peake, and T. Dabelsteen. 1999. Do female great tits (*Parus major*) assess males by eavesdropping? A field study using interactive song playback. *Proc. R. Soc. Lond. B* 266:1305–1309.

Otter, K., L. M. Ratcliffe, D. Michaud, and P. T. Boag. 1998. Do female black-capped chickadees prefer high-ranking males as extra-pair partners? *Behav. Ecol. Sociobiol.* 43:25–36.

Otter, K. A., I. R. K. Stewart, P. K. McGregor, A. M. R. Terry, T. Dabelsteen, and T. Burke. 2001. Extra-pair paternity among Great Tits *Parus major* following manipulation of male signals. *J. Avian Biol.* 32:338–344.

Owens, I. P. F., and I. R. Hartley. 1991. "Trojan sparrows": evolutionary consequences of dishonest invasion for the badges-of-status model. *Am. Nat.* 138:1187–1205.

Owings, D. H., and D. F. Hennessy. 1984. The importance of variation in sciurid visual and vocal communication. Pp. 169–200 in J. O. Murie and G. R. Michener, eds., *The Biology of Ground-dwelling Squirrels*. Univ. Nebraska Press, Lincoln.

Palokangas, P., R. V. Alatalo, and E. Korpimäki. 1992. Female choice in the kestrel under different availability of mating options. *Anim. Behav.* 43:659–665.

Parker, G. A., and M. R. Macnair. 1979. Models of parent-offspring conflict. IV. Suppression: evolutionary retaliation by the parent. *Anim. Behav.* 27:1210–1235.

Parker, G. A., N. J. Royle, and I. R. Hartley. 2002. Begging scrambles with unequal chicks: interactions between need and competitive ability. *Ecology Letters* 5: 206–215.

Pärt, T., and A. Qvarnström. 1997. Badge size in collared flycatchers predicts outcome of male competition over territories. *Anim. Behav.* 54:893–899.

Passmore, N. I., and S. R. Telford. 1983. Random mating by size and age of males in the painted reed frog, *Hyperolius marmoratus*. *S. Afr. J. Sci.* 79:353–355.

Payne, R. B., and K. Payne. 1977. Social organization and mating success in local song populations of village indigobirds, *Vidua chalybeata*. *Z. Tierpsychol.* 45:113–173.

Peake, T. M. In press. Eavesdropping in communication networks. In P. K. McGregor, ed., *Animal Communication Networks*. Cambridge Univ. Press, Cambridge.

Peake, T. M., and P. K. McGregor. 2004. Information and aggression in fishes. *Learning and Behavior* 32:114–121.

Peake, T. M., A. M. R. Terry, P. K. McGregor, and T. Dabelsteen. 2001. Male great tits eavesdrop on simulated male-to-male vocal interactions. *Proc. R. Soc. Lond. B* 268:1183–1187.

Peake, T. M., A. M. R. Terry, P. K. McGregor, and T. Dabelsteen. 2002. Do great tits assess rivals by combining direct experience with information gathered by eavesdropping? *Proc. R. Soc. Lond. B* 269:1925–1929.

Peters, A. 2000. Testosterone treatment is immunosuppressive in superb fairy-wrens, yet free-living males with high testosterone are more immunocompetent. *Proc. R. Soc. Lond. B* 267:883–889.

Petrie, M. 1994. Improved growth and survival of offspring of peacocks with more elaborate trains. *Nature* 371:598–599.

Petrie, M., and T. Halliday. 1994. Experimental and natural changes in the peacock's (*Pavo cristatus*) train can affect mating success. *Behav. Ecol. Sociobiol.* 35:213–217.

Petrie, M., T. Halliday, and C. Sanders. 1991. Peahens prefer peacocks with elaborate trains. *Anim. Behav.* 41:323–331.

Podos, J. 1996. Motor constraints on vocal development in a songbird. *Anim. Behav.* 51:1061–1070.

Podos, J. 1997. A performance constraint on the evolution of trilled vocalizations in a songbird family (Passeriformes: Emberizidae). *Evolution* 51:537–551.

Podos, J., and S. Nowicki. 2004. Performance limits on birdsong production. Pp. 318–341 in P. Marler and H. Slabbekoorn, eds., *Nature's Musicians: The Science of Birdsong*. Academic Press, New York.

Podos, J., S. Peters, T. Rudnicky, P. Marler, and S. Nowicki. 1992. The organization of song repertoires in song sparrows: themes and variations. *Ethology* 90:89–106.

Podos, J., J. K. Sherer, S. Peters, and S. Nowicki. 1995. Ontogeny of vocal tract movements during song production in song sparrows. *Anim. Behav.* 50:1287–1296.

Podos, J., J. A. Southall, and M. R. Rossi-Santos. 2004. Vocal mechanics in Darwin's finches: correlation of beak gape and song frequency. *J. Exp. Biol.* 207:607–619.

Poiani, A., A. R. Goldsmith, and M. R. Evans. 2000. Ectoparasites of house sparrows (*Passer domesticus*): an experimental test of the immunocompetence handicap hypothesis and a new model. *Behav. Ecol. Sociobiol.* 47:230–242.

Pomiankowski, A. 1987. Sexual selection: the handicap mechanism does work—sometimes. *Proc. R. Soc. Lond. B* 231:123–145.

Pomiankowski, A., and Y. Iwasa. 1998. Handicap signaling: loud and true? *Evolution* 52:928–932.

Pomiankowski, A., Y. Iwasa, and S. Nee. 1991. The evolution of costly mate preferences. I. Fisher and biased mutation. *Evolution* 45:1422–1430.

Popp, J. W. 1987. Risk and effectiveness in the use of agonistic displays by American goldfinches. *Behaviour* 103:141–156.

Price, K. 1998. Benefits of begging for yellow-headed blackbird nestlings. *Anim. Behav.* 56:571–577.

Price, K., H. Harvey, and R. Ydenberg. 1996. Begging tactics of nestling yellow-headed blackbirds, *Xanthocephalus xanthocephalus*, in relation to need. *Anim. Behav.* 51:421–435.

Proctor, H. C. 1991. Courtship in the water mite *Neumania papillator*: males capitalize on female adaptations for predation. *Anim. Behav.* 42:589–598.

Proctor, H. C. 1992. Sensory exploitation and the evolution of male mating behaviour: a cladistic test using water mites (Acari: Parasitengona). *Anim. Behav.* 44:745–752.

Pryke, S. R., and S. Andersson. 2003. Carotenoid-based status signalling in red-shouldered widowbirds (*Euplectes axillaris*): epaulet size and redness affect captive and territorial competition. *Behav. Ecol. Sociobiol.* 53:393–401.

Pryke, S. R., S. Andersson, and M. J. Lawes. 2001. Sexual selection of multiple handicaps in the red-collared widowbird: female choice of tail length but not carotenoid display. *Evolution* 55:1452–1463.

Pryke, S. R., S. Andersson, M. J. Lawes, and S. E. Piper. 2002. Carotenoid status signaling in captive and wild red-collared widowbirds: independent effects of badge size and color. *Behav. Ecol.* 13:622–631.

Quillfeldt, P. 2002. Begging in the absence of sibling competition in Wilson's storm-petrels, *Oceanites oceanicus*. *Anim. Behav.* 64:579–587.

Qvarnström, A. 1997. Experimentally increased badge size increases male competition and reduces male parental care in the collared flycatcher. *Proc. R. Soc. Lond. B* 264:1225–1231.

Radesäter, T., S. Jakobsson, N. Andbjer, A. Bylin, and K. Nyström. 1987. Song rate and pair formation in the willow warbler, *Phylloscopus trochilus*. *Anim. Behav.* 35:1645–1651.

Rand, A. S., and M. J. Ryan. 1981. The adaptive significance of a complex vocal repertoire in a neotropical frog. *Z. Tierpsychol.* 57:209–214.

Reby, D., and K. McComb. 2003. Anatomical constraints generate honesty: acoustic cues to age and weight in the roars of red deer stags. *Anim. Behav.* 65:519–530.

Redondo, T., and F. Castro. 1992. Signalling of nutritional need by magpie nestlings. *Ethology* 92:193–204.

Regosin, J. V., and S. Pruett-Jones. 2001. Sexual selection and tail-length dimorphism in scissor-tailed flycatchers. *Auk* 118:167–175.

Reid, J. M., P. Arcese, A. L. E. V. Cassidy, S. M. Hiebert, J. N. M. Smith, P. K. Stoddard, A. B. Marr, and L. F. Keller. 2004. Song repertoire size predicts initial mating success in male song sparrows *Melospiza melodia*. *Anim. Behav.* 68:1055–1063.

Reid, M. L. 1987. Costliness and reliability in the singing vigour of Ipswich sparrows. *Anim. Behav.* 35:1735–1743.

Ricklefs, R. E. 1974. Energetics of reproduction in birds. *Proc. Nuttall Ornith. Club* 15:152–292.

Ricklefs, R. E. 1983. Avian postnatal development. Pp. 1–83 in D. S. Farner, J. R. King, and K. C. Parkes, eds., *Avian Biology*. Academic Press, New York.

Ricklefs, R. E., and S. Peters. 1979. Intraspecific variation in the growth rate of nestling European starlings. *Bird-Banding* 50:338–348.

Ricklefs, R. E., and S. Peters. 1981. Parental components of variance in growth rate and body size of nestling European starlings (*Sturnus vulgaris*) in eastern Pennsylvania. *Auk* 98:39–48.

Riede, T., and W. T. Fitch. 1999. Vocal tract length and acoustics of vocalizations in the domestic dog (*Canis familiaris*). *J. Exp. Biol.* 202:2859–2867.

Rinden, H., H. M. Lampe, T. Slagsvold, and Y. O. Espmark. 2000. Song quality does not indicate male parental abilities in the pied flycatcher *Ficedula hypoleuca*. *Behaviour* 137:809–823.

Ringsby, T. H., B.-E. Saether, and E. J. Solberg. 1998. Factors affecting juvenile survival in house sparrow *Passer domesticus*. *J. Avian Biol.* 29:241–247.

Robertson, J. G. M. 1986. Male territoriality, fighting and asssessment of fighting ability in the Australian frog *Uperoleia rugosa*. *Anim. Behav.* 34:763–772.

Robinson, S. R. 1981. Alarm communication in Belding's ground squirrels. *Z. Tierpsychol.* 56:150–168.

Rodd, F. H., K. A. Hughes, G. F. Grether, and C. T. Baril. 2002. A possible non-sexual origin of mate preference: are male guppies mimicking fruit? *Proc. R. Soc. Lond. B* 269:475–481.

Rodríguez-Gironés, M. A., P. A. Cotton, and A. Kacelnik. 1996. The evolution of begging: signaling and sibling competition. *Proc. Nat. Acad. Sci. USA* 93:14637–14641.

Rodríguez-Gironés, M. A., M. Enquist, and P. A. Cotton. 1998. Instability of signaling resolution models of parent-offspring conflict. *Proc. Nat. Acad. Sci. USA* 95:4453–4457.

Rodríguez-Gironés, M. A., M. Enquist, and M. Lachmann. 2001a. Role of begging and sibling competition in foraging strategies of nestlings. *Anim. Behav.* 61:733–745.

Rodríguez-Gironés, M. A., J. M. Zúñiga, and T. Redondo. 2001b. Effects of begging on growth rates of nestling chicks. *Behav. Ecol.* 12:269–274.

Rodríguez-Gironés, M. A., J. M. Zúñiga, and T. Redondo. 2002. Feeding experience and relative size modify the begging strategies of nestlings. *Behav. Ecol.* 13:782–785.

Rohwer, S. 1975. The social significance of avian winter plumage variability. *Evolution* 29:593–610.

Rohwer, S. 1977. Status signaling in Harris sparrows: some experiments in deception. *Behaviour* 61:107–129.

Rohwer, S. 1985. Dyed birds achieve higher social status than controls in Harris' sparrows. *Anim. Behav.* 33:1325–1331.

Rohwer, S., and P. W. Ewald. 1981. The cost of dominance and advantage of subordination in a badge signaling system. *Evolution* 35:441–454.

Rohwer, S., P. W. Ewald, and F. C. Rohwer. 1981. Variation in size, appearance, and dominance within and among the sex and age classes of Harris' sparrows. *J. Field Ornithol.* 52:291–303.

Romanes, G. J. 1883. *Animal Intelligence*. Kegan Paul, Trench & Co., London.

Ros, A. F. H., T. G. G. Groothius, and V. Apanius. 1997. The relation among gonadal steroids, immunocompetence, body mass, and behavior in young black-headed gulls (*Larus ridibundus*). *Am. Nat.* 150:201–219.

Røskaft, T., T. Järvi, M. Bakken, C. Bech, and R. E. Reinertsen. 1986. The relationship between social status and resting metabolic rate in great tits (*Parus major*) and pied flycatchers (*Ficedula hypoleuca*). *Anim. Behav.* 34:838–842.

Rowe, L. V., M. R. Evans, and K. L. Buchanan. 2001. The function and evolution of the tail streamer in hirundines. *Behav. Ecol.* 12:157–163.

Royle, N. J., I. R. Hartley, and G. A. Parker. 2002. Begging for control: when are offspring solicitation behaviours honest? *Trends Ecol. Evol.* 17:434–440.

Ruff, M. D., W. M. Reid, and J. K. Johnson. 1974. Lowered blood carotenoid levels in chickens infected with coccidia. *Poultry Science* 53:1801–1809.

Rundle, H. D., and D. Schluter. 1998. Reinforcement of stickleback mate preferences: sympatry breeds contempt. *Evolution* 52:200–208.

Russow, L.-M. 1986. Deception: a philosophical perspective. Pp. 41–65 in R. W. Mitchell and N. S. Thompson, eds., *Deception: Perspectives on Human and Nonhuman Deceit*. SUNY Press, Albany.

Ryan, M. J. 1980. Female choice in a neotropical frog. *Science* 209:523–525.

Ryan, M. J. 1983. Sexual selection and communication in a neotropical frog, *Physalaemus pustulosus*. *Evolution* 37:261–272.

Ryan, M. J. 1985a. Energetic efficiency of vocalization by the frog *Physalaemus pustulosus*. *J. Exp. Biol.* 116:47–52.

Ryan, M. J. 1985b. *The Túngara Frog*. Univ. Chicago Press, Chicago.

Ryan, M. J. 1998. Sexual selection, receiver biases, and the evolution of sex differences. *Science* 281:1999–2003.

Ryan, M. J., J. H. Fox, W. Wilczynski, and A. S. Rand. 1990. Sexual selection for sensory exploitation in the frog *Physalaemus pustulosus*. *Nature* 343:66–67.

Ryan, M. J., and A. S. Rand. 1993. Sexual selection and signal evolution: the ghost of biases past. *Phil. Trans. Roy. Soc. Lond. B* 340:187–195.

Ryan, M. J., M. D. Tuttle, and A. S. Rand. 1982. Bat predation and sexual advertisement in a neotropical anuran. *Am. Nat.* 119:136–139.

Saetre, G.-P., T. Moum, S. Bures, M. Král, M. Adamjan, and J. Moreno. 1997. A sexually selected character displacement in flycatchers reinforces premating isolation. *Nature* 387:589–592.

Safran, R. J., and K. J. McGraw. 2004. Plumage coloration, not length or symmetry of tail-streamers, is a sexually selected trait in North American barn swallows. *Behav. Ecol.* 15:455–461.

Saino, N., R. Ambrosini, R. Martinelli, P. Ninni, and A. P. Møller. 2003. Gape coloration reliably reflects immunocompetence of barn swallow (*Hirundo rustica*) nestlings. *Behav. Ecol.* 14:16–22.

Saino, N., J. J. Cuervo, P. Ninni, F. de Lope, and A. P. Møller. 1997a. Haematocrit correlates with tail ornament size in three populations of the barn swallow (*Hirundo rustica*). *Functional Ecology* 11:604–610.

Saino, N., A. P. Møller, and A. M. Bolzern. 1995. Testosterone effects on the immune system and parasite infestations in the barn swallow (*Hirundo rustica*): an experimental test of the immunocompetence hypothesis. *Behav. Ecol.* 6:397–404.

Saino, N., P. Ninni, S. Calza, R. Martinelli, F. De Bernardi, and A. P. Møller. 2000. Better red than dead: carotenoid-based mouth coloration reveals infection in barn swallow nestlings. *Proc. R. Soc. Lond. B* 267:757–761.

Saino, N., C. R. Primmer, H. Ellegren, and A. P. Møller. 1997b. An experimental study of paternity and tail ornamentation in the barn swallow (*Hirundo rustica*). *Evolution* 51:562–570.

Sakaluk, S. K., and J. J. Belwood. 1984. Gecko phonotaxis to cricket calling song: a case of satellite predation. *Anim. Behav.* 32:659–662.

Schwagmeyer, P. L. 1980. Alarm calling behavior of the thirteen-lined ground squirrel, *Spermophilus tridecemlineatus*. *Behav. Ecol. Sociobiol.* 7:195–200.

Schwagmeyer, P. L., and D. W. Foltz. 1990. Factors affecting the outcome of sperm competition in thirteen-lined ground squirrels. *Anim. Behav.* 39:156–162.

Scott, G. W., and J. M. Deag. 1998. Blue tit (*Parus caeruleus*) agonistic displays: a reappraisal. *Behaviour* 135:665–691.

Searcy, W. A. 1979. Sexual selection and body size in male red-winged blackbirds. *Evolution* 33:649–661.

Searcy, W. A. 1984. Song repertoire size and female preferences in song sparrows. *Behav. Ecol. Sociobiol.* 14:281–286.

Searcy, W. A., and M. Andersson. 1986. Sexual selection and the evolution of song. *Ann. Rev. Ecol. Syst.* 17:507–533.

Searcy, W. A., and P. Marler. 1981. A test for responsiveness to song structure and programming in female sparrows. *Science* 213:926–928.

Searcy, W. A., and S. Nowicki. 2000. Male-male competition and female choice in the evolution of vocal signaling. Pp. 301–315 in Y. Espmark, T. Amundsen, and G. Rosenqvist, eds., *Animal Signals: Signalling and Signal Design in Animal Communication*. Tapir Academic Press, Trondheim, Norway.

Searcy, W. A., S. Nowicki, and M. Hughes. 1997. The response of male and female song sparrows to geographic variation in song. *Condor* 99:651–657.

Searcy, W. A., S. Nowicki, M. Hughes, and S. Peters. 2002. Geographic song discrimination in relation to dispersal distances in song sparrows. *Am. Nat.* 159:221–230.

Searcy, W. A., S. Nowicki, and S. Peters. 2003. Phonology and geographic song discrimination in song sparrows. *Ethology* 109:23–35.

Searcy, W. A., S. Peters, and S. Nowicki. 2004. Effects of early nutrition on growth rate and adult size in song sparrows *Melospiza melodia*. *J. Avian Biol.* 35:269–279.

Searcy, W. A., and K. Yasukawa. 1990. Use of the song repertoire in intersexual and intrasexual contexts by male red-winged blackbirds. *Behav. Ecol. Sociobiol.* 27:123–128.

Semler, D. E. 1971. Some aspects of adaptation in a polymorphism for breeding colours in the threespine stickleback (*Gasterosteus aculeatus*). *J. Zool.* 165:291–302.

Senar, J. C., M. Camerino, J. L. Copete, and N. B. Metcalfe. 1993. Variation in black bib of the Eurasian siskin (*Carduelis spinus*) and its role as a reliable badge of dominance. *Auk* 110:924–927.

Senar, J. C., V. Polo, F. Uribe, and M. Camerino. 2000. Status signalling, metabolic rate and body mass in the siskin: the cost of being a subordinate. *Anim. Behav.* 59:103–110.

Seyfarth, R. M., and D. L. Cheney. 1980. The ontogeny of vervet monkey alarm calling behavior: a preliminary report. *Z. Tierpsychol.* 54:37–56.

Seyfarth, R. M., and D. L. Cheney. 1986. Vocal development in vervet monkeys. *Anim. Behav.* 34:1640–1658.

Seyfarth, R. M., and D. L. Cheney. 2003. Signalers and receivers in animal communication. *Ann. Rev. Psychol.* 54:145–173.

Seyfarth, R. M., D. L. Cheney, and P. Marler. 1980. Vervet monkey alarm calls: semantic communication in a free-ranging primate. *Anim. Behav.* 28:1070–1094.

Sherman, P. W. 1977. Nepotism and the evolution of alarm calls. *Science* 197:1246–1253.

Sherman, P. W. 1980. The meaning of nepotism. *Am. Nat.* 116:604–606.

Sherman, P. W. 1985. Alarm calls of Belding's ground squirrels to aerial predators: nepotism or self-preservation? *Behav. Ecol. Sociobiol.* 17:313–323.

Shields, W. M. 1980. Ground squirrel alarm calls: nepotism or parental care? *Am. Nat.* 116:599–603.

Shykoff, J. A., and A. Widmer. 1996. Parasites and carotenoid-based signal intensity: how general should the relationship be? *Naturwissenschaften* 83:113–121.

Silk, J. B., E. Kaldor, and R. Boyd. 2000. Cheap talk when interests conflict. *Anim. Behav.* 59:423–432.

Simpson, M. J. A. 1968. The display of the Siamese fighting fish, *Betta splendens*. *Anim. Behav. Monogr.* 1:1–73.

Slagsvold, T., and J. T. Lifjeld. 1985. Variation in plumage colour of the great tit *Parus major* in relation to habitat, season and food. *J. Zool.* 206:321–328.

Slater, P. J. B., F. A. Clements, and D. J. Goodfellow. 1984. Local and regional variations in chaffinch song and the question of dialects. *Behaviour* 88:76–97.

Slater, P. J. B., L. A. Eales, and N. S. Clayton. 1988. Song learning in zebra finches (*Taeniopygia guttata*): progress and prospects. *Adv. Study Behav.* 18:1–34.

Smart, J. L. 1986. Undernutrition, learning and memory: review of experimental studies. Pp. 74–78 in T. G. Taylor and N. K. Jenkins, eds. *Proc. 13th Congress of Nutrition.* Libbey, London.

Smiseth, P. T., and S.-H. Lorentsen. 2001. Begging and parent-offspring conflict in grey seals. *Anim. Behav.* 62:273–279.

Smith, H. G., and R. Montgomerie. 1991a. Nestling American robins compete with siblings by begging. *Behav. Ecol. Sociobiol.* 29:307–312.

Smith, H. G., and R. Montgomerie. 1991b. Sexual selection and the tail ornaments of North American barn swallows. *Behav. Ecol. Sociobiol.* 28:195–201.

Smith, H. G., R. Montgomerie, T. Pöldmaa, B. N. White, and P. T. Boag. 1991. DNA fingerprinting reveals relation between tail ornaments and cuckoldry in barn swallows, *Hirundo rustica. Behav. Ecol.* 2:90–98.

Smith, S. F. 1978. Alarm calls, their origin and use in *Eutamias sonomae. J. Mamm.* 59:888–893.

Smythe, N. 1970. On the existence of "pursuit invitation" signals in mammals. *Am. Nat.* 104:491–494.

Soha, J. A., D. A. Nelson, and P. G. Parker. 2004. Genetic analysis of song dialect populations in Puget Sound white-crowned sparrows. *Behav. Ecol.* 15:636–646.

Sohrabji, F., E. J. Nordeen, and K. W. Nordeen. 1990. Selective impairment of song learning following lesions of a forebrain nucleus in the juvenile zebra finch. *Behav. Neural Biol.* 53:51–63.

Spencer, K. A., K. L. Buchanan, A. R. Goldsmith, and C. K. Catchpole. 2004. Developmental stress, social rank and song complexity in the European starling (*Sturnus vulgaris*). *Proc. R. Soc. Lond. B* 271:S121-S123.

Stamps, J., A. Clark, P. Arrowood, and B. Kus. 1989. Begging behaviour in budgerigars. *Ethology* 81:177–192.

Steger, R., and R. L. Caldwell. 1983. Intraspecific deception by bluffing: a defense strategy of newly molted stomatopods (Arthropoda: Crustacea). *Science* 221:558–560.

Stokes, A. W. 1962. Agonistic behaviour among blue tits at a winter feeding station. *Behaviour* 19:118–138.

Stokes, A. W. 1971. Parental and courtship feeding in red jungle fowl. *Auk* 88:21–29.

Struhsaker, T. T. 1967. Auditory communication among vervet monkeys (*Cercopithecus aethiops*). Pages 281–324 in S. A. Altman, ed., Social communication among primates. Univ. Chicago Press, Chicago.

Stutchbury, B. J. M., and E. S. Morton. 2001. *Behavioral Ecology of Tropical Birds*. Academic Press, New York.

Sullivan, K. A. 1988. Age-specific profitability and prey choice. *Anim. Behav.* 36:613–615.

Suthers, R. A., and F. Goller. 1997. Motor correlates of vocal diversity in songbirds. *Current Ornithology* 14:235–288.

Számadó, S. 2000. Cheating as a mixed strategy in a simple model of aggressive communication. *Anim. Behav.* 59:221–230.

Székely, T., C. K. Catchpole, A. DeVoogd, Z. Marchl, and T. J. DeVoogd. 1996. Evolutionary changes in a song control area of the brain (HVC) are associated with evolutionary changes in song repertoire among European warblers (Sylviidae). *Proc. Roy. Soc. Lond. B* 263:607–610.

Tamura, M. 1995. Postcopulatory mate guarding by vocalization in the Formosan squirrel. *Behav. Ecol. Sociobiol.* 36:377–386.

Terry, A. M. R., and R. Lachlan. In press. Communication networks in a virtual world. in P. K. McGregor, ed., *Animal Communication Networks*. Cambridge Univ. Press, Cambridge.

Thompson, C. W., N. Hillgarth, M. Leu, and H. E. McClure. 1997. High parasite load in house finches (*Carpodacus mexicanus*) is correlated with reduced expression of a sexually selected trait. *Am. Nat.* 149:270–294.

Tinbergen, N. 1951. *The Study of Instinct*. Oxford Univ. Press, Oxford.

Tinbergen, N. 1964. The evolution of signaling devices. Pp. 206–230 in W. Etkin, ed., *Social Behavior and Organization Among Vertebrates*. Univ. Chicago Press, Chicago.

Todt, D. 1981. On functions of vocal matching: effects of counter-replies on song post choice and singing. *Z. Tierpsychol.* 57:73–93.

Todt, D., and M. Naguib. 2000. Vocal interactions in birds: the use of song as a model in communication. Pp. 247–296 in P. J. B. Slater, J. S. Rosenblatt, C. T. Snowdon, and T. J. Roper, eds., *Advances in the Study of Behavior*. Academic Press, New York.

Trivers, R. L. 1972. Parental investment and sexual selection. Pp. 136–179 in B. Campbell, ed., *Sexual Selection and the Descent of Man*. Aldine, Chicago.

Tyczkowski, J. K., P. B. Hamilton, and M. D. Ruff. 1991. Altered metabolism of carotenoids during pale-bird syndrome in chickens infected with *Eimeria acervulina*. *Poultry Science* 70:2074–2081.

Vallet, E., I. Beme, and M. Kreutzer. 1998. Two-note syllables in canary songs elicit high levels of sexual display. *Anim. Behav.* 55:291–297.

Vallet, E., and M. Kreutzer. 1995. Female canaries are sexually responsive to special song phrases. *Anim. Behav.* 49:1603–1610.

van Dommelen, W. A., and B. H. Moxness. 1995. Acoustic parameters in speaker height and weight identification: sex-specific behaviour. *Language and Speech* 38:267–287.

van Krunkelsven, E., J. Dupain, L. Van Elsacker, and R. F. Verheyn. 1996. Food calling by captive bonobos (*Pan paniscus*): an experiment. *Int. J. Primatol.* 17:207–217.

van Rhijn, J. G. 1980. Communication by agonistic displays: a discussion. *Behaviour* 74:284–293.

Vehrencamp, S. L. 2000. Handicap, index, and conventional signal elements of bird song. Pp. 277–300 in Y. Espmark, T. Amundsen, and G. Rosenqvist, eds., *Animal Signals: Signalling and Signal Design in Animal Communication.* Tapir Academic Press, Trondheim, Norway.

Veiga, J. P. 1993. Badge size, phenotypic quality, and reproductive success in the house sparrow: a study on honest advertisement. *Evolution* 47:1161–1170.

Veiga, J. P. 1995. Honest signaling and the survival cost of badges in the house sparrow. *Evolution* 49:570–572.

Veiga, J. P., and M. Puerta. 1996. Nutritional constraints determine the expression of a sexual trait in the house sparrow, *Passer domesticus. Proc. R. Soc. Lond. B* 263:229–234.

von Haartman, L. 1953. Was reizt den Trauerfliegenschnäpper (*Muscicapa hypoleuca*) zu füttern? *Die Vogelwarte* 16:157–164.

von Schantz, T., S. Bensch, M. Grahn, D. Hasselquist, and H. Witzell. 1999. Good genes, oxidative stress and condition-dependent sexual signals. *Proc. R. Soc. Lond. B* 266:1–12.

Waas, J. R. 1991. The risks and benefits of signalling aggressive motivation: a study of cave-dwelling little blue penguins. *Behav. Ecol. Sociobiol.* 29:139–146.

Wagner, W. E. 1989. Fighting, assessment, and frequency alteration in Blanchard's cricket frog. *Behav. Ecol. Sociobiol.* 25:429–436.

Wagner, W. E. 1992. Deceptive or honest signalling of fighting ability? A test of alternative hypotheses for the function of changes in call dominant frequency by male cricket frogs. *Anim. Behav.* 44:449–462.

Walther, F. R. 1969. Flight behaviour and avoidance of predators in Thomson's gazelle (*Gazella thomsoni* Guenther 1884). *Behaviour* 34:184–221.

Ward, S., H. Lampe, and P. J. B. Slater. 2004. Singing is not energetically demanding for pied flycatchers, *Ficedula hypoleuca. Behav. Ecol.* 15:477–484.

Ward, S., J. R. Speakman, and P. J. B. Slater. 2003. The energy cost of song in the canary, *Serinus canaria. Anim. Behav.* 66:893–902.

Wauters, A.-M., and M.-A. Richard-Yris. 2002. Mutual influence of the maternal hen's food calling and feeding behavior on the behavior of her chicks. *Dev. Psychobiol.* 41:25–36.

Wauters, A.-M., and M.-A. Richard-Yris. 2003. Maternal food calling in domestic hens: influence of feeding context. *C. R. Biologies* 326:677–686.

Wauters, A.-M., M.-A. Richard-Yris, J. S. Pierre, C. Lunel, and J. P. Richard. 1999. Influence of chicks and food quality on food calling in broody domestic hens. *Behaviour* 136:919–933.

Weary, D. M., and D. Fraser. 1995. Calling by domestic piglets: reliable signals of need? *Anim. Behav.* 50:1047–1055.

Weary, D. M., and J. R. Krebs. 1992. Great tits classify songs by individual voice characteristics. *Anim. Behav.* 43:283–287.

Weary, D. M., G. L. Lawson, and B. K. Thompson. 1996. Sows show stronger responses to isolation calls of piglets associated with greater levels of piglet need. *Anim. Behav.* 52:1247–1253.

Weatherhead, P. J., K. J. Metz, G. F. Bennett, and R. E. Irwin. 1993. Parasite faunas, testosterone and secondary sexual traits in male red-winged blackbirds. *Behav. Ecol. Sociobiol.* 33:13–23.

Weathers, W. W., and K. A. Sullivan. 1989. Juvenile foraging proficiency, parental effort, and avian reproductive success. *Ecol. Monogr.* 59:223–246.

Welling, P. P., S. O. Rytkönen, K. T. Koivula, and M. I. Orell. 1997. Song rate correlates with paternal care and survival in willow tits: advertisement of male quality? *Behaviour* 134:891–904.

Wells, K. D. 1977. The social behaviour of anuran amphibians. *Anim. Behav.* 25:666–693.

Wells, K. D., and T. L. Taigen. 1986. The effect of social interactions on calling energetics in the gray treefrog (*Hyla versicolor*). *Behav. Ecol. Sociobiol.* 19:9–18.

West, M. J., and A. P. King. 1980. Enriching cowbird song by social deprivation. *J. Comp. Physiol. Psych.* 94:263–270.

West, M. J., A. P. King, and D. H. Eastzer. 1981. Validating the female bioassay of cowbird song: relating differences in song potency to mating success. *Anim. Behav.* 29:490–501.

West, M. J., A. P. King, D. H. Eastzer, and J. E. R. Staddon. 1979. A bioassay of isolate cowbird song. *J. Comp. Physiol. Psych.* 93:124–133.

West, M. J., A. P. King, and T. J. Harrocks. 1983. Cultural transmission of cowbird song (*Molothrus ater*): measuring its development and outcome. *J. Comp. Psych.* 97:327–337.

West-Eberhard, M. J. 1979. Sexual selection, competition, and evolution. *Proc. Am. Phil. Soc.* 123:222–234.

Westneat, M. W., J. H. Long, W. Hoese, and S. Nowicki. 1993. Kinematics of birdsong: functional correlation of cranial movements and acoustic features in sparrows. *J. Exp. Biol.* 182:147–171.

Whiten, A., and R. W. Byrne. 1988. Tactical deception in primates. *Behav. Brain Sci.* 11:233–273.

Whitfield, D. P. 1987. Plumage variability, status signalling and individual recognition in avian flocks. *Trends Ecol. Evol.* 2:13–18.

Wiley, R. H. 1994. Errors, exaggeration, and deception in animal communication. Pp. 157–189 in L. A. Real, ed., *Behavioral Mechanisms in Evolutionary Ecology*. University Chicago Press, Chicago.

Williams, G. C. 1966. Adaptation and Natural Selection. Princeton Univ. Press, Princeton, NJ.

Williams, H. W. 1969. Vocal behavior of adult California quail. *Auk* 86:631–659.

Williams, H. W., A. W. Stokes, and J. C. Wallen. 1968. The food call and display of the bobwhite quail (*Colinus virginianus*). *Auk* 85:464–476.

Wilson, E. O. 1975. *Sociobiology*. Harvard Univ. Press, Cambridge, MA.

Wolfenbarger, L. L. 1999. Is red coloration of male northern cardinals beneficial during the nonbreeding season?: a test of status signaling. *Condor* 101:655–663.

Woodland, D. J., Z. Jaafar, and M. L. Knight. 1980. The "pursuit deterrent" function of alarm signals. *Am. Nat.* 115:748–753.

Wright, J., C. Hinde, I. Fazey, and C. Both. 2002. Begging signals more than just short-term need: cryptic effects of brood size in the pied flycatcher (*Ficedula hypoleuca*). *Behav. Ecol. Sociobiol.* 52:74–83.

Wynne-Edwards, V. C. 1962. *Animal Dispersion in Relation to Social Behaviour.* Oliver and Boyd, Edinburgh.

Yasukawa, K., J. L. Blank, and C. B. Patterson. 1980. Song repertoires and sexual selection in the red-winged blackbird. *Behav. Ecol. Sociobiol.* 7:233–238.

Zahavi, A. 1975. Mate selection—a selection for a handicap. *J. Theor. Biol.* 53:205–214.

Zahavi, A. 1977a. The cost of honesty (further remarks on the handicap principle). *J. Theor. Biol.* 67:603–605.

Zahavi, A. 1977b. Reliability in communications systems and the evolution of altruism. Pp. 253–259 in B. Stonehouse and C. M. Perrins, eds., *Evolutionary Ecology.* MacMillan, London.

Zahavi, A. 1987. The theory of signal selection and some of its implications. Pp. 305–327 in V. P. Delfino, ed., *International Symposium of Biological Evolution.* Adriatic Editrice, Bari, Italy.

Zahavi, A., and A. Zahavi. 1997. *The Handicap Principle: A Missing Piece of Darwin's Puzzle.* Oxford Univ. Press, Oxford.

Zink, R. M., and G. F. Barrowclough. 1984. Allozymes and song dialects: a reassessment. *Evolution* 38:444–448.

Zink, R. M., and D. L. Dittmann. 1993. Gene flow, refugia, and evolution of geographic variation in the song sparrow (*Melospiza melodia*). *Evolution* 47:717–729.

Zuberbühler, K., D. Jenny, and R. Bshary. 1999. The predator deterrence function of primate alarm calls. *Ethology* 105:477–490.

Zweifel, R. G. 1968. Effects of temperature, body size, and hybridization on mating calls of toads, *Bufo a. americanus* and *Bufo woodhousii fowleri. Copeia* 1968:269–285.

Author Index

Subject Index

ᛁ

DATE DUE